Stefan Peter Müller

Compression of an array of similar crash test simulation results

Logos Verlag Berlin

Bibliografische Information der Deutschen Nationalbibliothek

Die Deutsche Nationalbibliothek verzeichnet diese Publikation in der Deutschen Nationalbibliografie; detaillierte bibliografische Daten sind im Internet über http://dnb.d-nb.de abrufbar.

The model has been developed by The National Crash Analysis Center (NCAC) of The George Washington University under a contract with the FHWA and NHTSA of the US DOT.

This work was submitted as a dissertation to the Humboldt Universität zu Berlin.

ISBN 978-3-8325-5444-6

Logos Verlag Berlin GmbH
Georg-Knorr-Str. 4, Geb. 10,
12681 Berlin
Tel.:   +49 (0)30 / 42 85 10 90
Fax: +49 (0)30 / 42 85 10 92
http://www.logos-verlag.de

To my grandfather Christoph.

Zusammenfassung

In der vorliegenden Arbeit wird eine Technik zur verlustbehafteten Kompression von Scharen an ähnlichen Simulationsergebnissen beschrieben. Die entwickelten Techniken nutzen vorhandene Ähnlichkeiten zwischen Simulationsergebnissen aus, um eine effizientere Kompression zu erreichen als für den Fall, dass jedes Simulationsmodell für sich behandelt werden würde. Der Fokus wurde hierbei auf Crashtest Simulationsergebnisse gelegt, die in großer Zahl in der Fahrzeugentwicklung generiert werden. Für den vorgestellten Kompressionsansatz, wird das Fahrzeugmodell in durch die Simulationsergebnisse definierten Bauteile zerlegt. Eine Crashtest Simulationsdatei besteht zu einem Großteil aus zeitabhängigen Knoten- und Elementvariablen. Durch die Zerlegung des Modells in Bauteile und das Wissen, welche Bauteile in welchen Simulationsergebnissen vorkommen, können die zeitabhängigen Variablen eines Bauteils für alle Simulationsergebnisse extrahieren werden. Pro Variable und pro Bauteil wird eine Datenmatrix angelegt, in der jede Spalte einem Zeitschritt eines Simulationsergebnisses entspricht. Auf diese Matrizen wird die Predictive Principal Component Analysis (PPCA) Methode angewendet. Die PPCA Methode führt zunächst eine PCA auf einer Datenmatrix aus. Über einen Optimierungsprozess wird die Anzahl der Hauptkomponenten bestimmt, die zur Rekonstruktion der Daten verwendet werden sollen. Die Rekonstruktion wird im Anschluss als Vorhersage der Datenmatrix verwendet. Die PPCA Methoden können sowohl als Offline Verfahren direkt auf eine ganze Schar an Simulationsergebnissen angewandt werden, wie auch als Online Verfahren, das ein nachträgliches Hinzufügen von Simulationsergebnissen erlaubt. Sowohl in der Online als auch im Offline Variante treten mehrere frei wählbare Parameter auf, die einen erheblichen Einfluss auf die Kompressionsgüte und -geschwindigkeit haben können. Diese werden sowohl theoretisch als auch empirisch anhand von 3 Testdatensätzen untersucht und mit dem State-of-the-art Kompressionsprogramm FEMZIP verglichen. Zudem wurde die Kennzahl Lernfaktor eingeführt, mit der man Klassifizieren kann, ob ein Verfahren davon profitiert, dass nicht nur eins, sondern eine Vielzahl an Simulationen komprimiert wird. Das Ergebnis der PPCA Vorhersage ist ein Residuen-Matrix, die die gleiche Dimension wie die ursprüngliche Datenmatrix besitzt. Zur Kompression der Residuen-Matrix wurde der induced Markov chain (iMc) Kodierer angewandt. Der iMc Kodierer ist der erste Kodierer, der die Topologie des Gitters als Nebeninformation ausnutzt und gleichzeitig in Fällen eines großen Wertebereiches angewandt werden kann. Neben der Beschreibung der praktischen Umsetzung des iMc Kodierers, werden die induced Markov chains als zugrundeliegende Datenmodell hergeleitet. Auf Basis des Datenmodells kann einerseits die Entropy des Datensatzes bestimmt werden, was für die Laufzeit des Optimierungsschritt der PPCA Methoden vorteilhaft ist. Andererseits kann die Qualität der iMc Kodierung theoretisch untersucht werden. Es wird zudem gezeigt, dass die stationäre Verteilung der zugrundeliegenden Markov Kette sich für alle praktischen Anwendungsfälle nur geringfügig von der Verteilung, die durch die relativen Häufigkeiten induziert wird, unterscheidet. Des Weiteren kann die Abweichung allein auf Basis der Topologie des Gitters abgeschätzt werden. Die Evaluation des iMc Kodierers erfolgt im Vergleich zu den Kodierer Rice und zlib auf den Residuen-Matrizen, die aus den PPCA Online und Offline entstehen. Dabei erzielt die Kombination des PPCA Offline Verfahrens mit dem iMc Kodierer für alle Benchmark Datensätze die besten Ergebnisse. Für das PPCA Online-Verfahren empfehlen wir die Nutzung des Rice Kodierers.

Abstract

This thesis describes a technique for lossy compression of an array of similar simulation results. The developed techniques use similarities between simulation results to achieve a more efficient compression than if each simulation model would be treated separately. We focus on crash test simulation results, which are generated in large numbers in vehicle development. For our compression approach, we decompose the vehicle model into parts that are defined in the simulation results. A crash test simulation file consists to a large extent of time-dependent node and element variables. By decomposing the model into components and knowing which components occur in which simulation results, the time-dependent variables of a part can be extracted for all simulation results. A data matrix is created per variable and per component, in which each column corresponds to a time step of a simulation result. The Predictive Principal Component Analysis (PPCA) method is applied to these matrices. The PPCA method first executes a PCA on a data matrix. A specialized optimization process is used to determine the number of principal components to be used to reconstruct the matrix. We use the reconstruction of the dimensionality reduction as a prediction of the data matrix. The PPCA methods can be applied both as offline method directly to a whole set of simulation results and as a online method, which allows to add simulation results afterwards. In both the online and offline variants, several freely selectable parameters occur which are crucial for the compression quality and speed. These parameters are investigated both theoretically and empirically on the basis of 3 test data sets. The results are compared with the state-of-the-art compression tool FEMZIP. In addition, the learning factor was introduced, which can be used to classify if a procedure benefits from the fact that a large amount of simulations are compressed rather than just one. The result of the PPCA prediction is a residual matrix that has the same dimension as the original data matrix. The induced Markov chain (iMc) encoder was used to compress the residual matrix. The iMc encoder is the first encoder that uses the topology of the grid as side information and - at the same time - can be applied in cases of a big alphabet. Besides the description of the practical implementation of the iMc encoder, the induced Markov chains are derived as the underlying data model. On the one hand, the entropy of the data set can be determined on the basis of the data model, which is advantageous for the runtime of the optimization step of the PPCA methods. On the other hand, the quality of the iMc coding can be investigated theoretically. It is shown that the stationary distribution of the underlying Markov chain for all practical applications differs only slightly from the distribution that is induced by the relative frequencies. In addition, the deviation can be estimated only on the basis of the topology of the grid. The evaluation of the iMc encoder is carried out in comparison to the encoders Rice and zlib on the residual matrices, which arise from an application of the PPCA Online and Offline methods on our benchmark data sets. The combination of the PPCA Offline method with the iMc encoder achieved the best results for all data sets. For the PPCA Online method, the results are not clear, as there are cases where the iMc encoder performs best as well as cases where the Rice encoder performs best. Since the Rice encoder is faster than the iMc encoder, we recommend to use the Rice encoder for the PPCA Online method.

Contents

Chapter 1

Introduction

Today, big data is a buzzword for a collection of data sets that is too large and too complex to process using traditional data processing algorithms. Since large automotive companies simulate several petabytes of data, the numerical simulations themselves represent a major data problem.

The focus of the application field examined in this thesis is the simulation of car crashes which dates back to the early 1980s [58]. This research area is of central importance for the development and quality management of automobile manufacturers, but at the same time leads to a constantly increasing demand for memory and bandwidth. Safety regulations and quality management require that part of the simulations be stored. The duration can range from a few months to several years. On the one hand, this leads to a large number of stored data records. On the other hand, the simulation results increase because the engineers want to improve the accuracy of the simulation results, e.g. by refining the model. The growing computing power makes this refinements possible.

A simulation data management system (SDMS) is usually applied to organize and handle a large number of simulation results. An SDMS enables the use of compression methods, that exploit redundancies between similar simulation results. An SDMS allows the definition of parts by connectivity and initial coordinates, which can only be stored once for a series of simulations instead of every time they occur in a simulation [118]. In addition, the time-dependent data of similar simulations, which make up the largest part of the data in the car crash result files, show high correlations among each other. It is, however, due to variations in the geometry, not trivial to exploit these correlations.

We have decided to duplicate nodes that occur in several components and to accept the overhead produced by this. Then, for each part individually, we apply a dimensionality reduction method to the information of all time steps for all available simulations that contain this part. Due to the decompression speed requirement, we further propose to perform a principal component analysis because its decompression can be accomplished by fast matrix operations and has the properties of random time step access and a progressive transmission. Since a simulation of a car crash takes several hours to a day, the time consumption of the compression is not crucial and an asymmetric compression method, in which the compression consumes more computational power, time, and memory, can be applied.

A disadvantage of the principal component analysis (PCA) - as with all dimensionality reduction methods - is that the bound of reconstruction error is not sharp [79]. But limiting the maximal absolute error is a key demand for the engineers to apply lossy data compression to simulation results. Due to strong restrictions of precision, it is almost impossible to apply a dimensionality reduction directly as a black box method and achieve a good compression rate. Therefore, we use the reconstruction of the PCA as an approximation for the initial data and calculate the residual between these two data sets. Moreover, we apply a lossy data compression on the residual and keep the low dimensional representation as side information, see [80]. We call this strategy Predictive Principal Component Analysis (PPCA). PPCA can be categorized as a prediction method, since we compress a residual instead of the original data set, see Section 3.3.2. Moreover, we want to quantize the principal components and the coefficients for both a better compression rate and to assure that the decompression on different machines generates identical results. Therefore, in the decompression phase, we use only integer arithmetic for matrix calculations. The proposed dimensionality reduction method only exploits directly the redundancies between simulations and time steps. But the data is usually defined on a finite element grid. Therefore, we investigate how to encode integers assigned to the nodes of a finite graph. The topological relation of neighbored nodes will be transformed into a value-based relation by calculating so-called transition probabilities. Combined with the relative frequencies of the node values, these probabilities form a Markov chain as the initial distribution. Therefore, we call our strategy induced Markov chains (iMc). Since each node is identified with a random variable, the iMc approach can be categorized as a special case of Bayesian networks. We will investigate the properties of the iMc especially its data compression properties. We prove that the iMc of connected parts is mean ergodic, and, therefore, the entropy as a measure of the amount of information is well defined. The encoding of Markov chains and their trajectories is a well known field in the information theory [121]. For a practical implementation, we determine a minimum span tree for the graph. For this tree the iMc statistics are determined, which are used as secondary information for our arithmetic encoder. We call this approach iMc encoder. Furthermore, we investigate alternative approaches to generate statistics for our arithmetic encoder.

In summary, when simulating a car crash we are often confronted with a high redundancy of the data, e.g. between time steps, neighbors and models. For a good compression rate it is crucial to eliminate these redundancies. This task can be tackled in two ways. The first is to predict the data so that the resulting variables are independent or at least nearly independent and encode the residual. The second is to find and exploit the remaining dependencies in the data. A prediction usually modifies the distribution of a data set. We will investigate a combination of these two approaches. For the prediction part, we will use the PPCA dimensionality reduction approach. Regarding the remaining dependencies, we propose the iMc as a two-pass universal encoder, which is based on coding and sending empirical data measurements rather than a coding method based on an a priori probability model.

1.1   Overview of proposed lossy compression method

In this section, we give a short overview on all steps of our compression method that is presented in this dissertation. We focus on the compression of time dependent coordinates and variables since the topology of the grid only needs to be stored once. Although, the PPCA can be combined with other encoders and the iMc encoder can be combined with different prediction methods, we see the combination of PPCA and iMc in several application fields as beneficial and propose it as a combined method.

We handle two cases. First, we consider cases where we have access to all simulation results we want to compress and second a case where simulations can be added after a first compression with all available simulation results. Therefore, the first situation is the initial stage of the second one. Furthermore, we assume that a user provides absolute precision for coordinates and variables. The reconstruction error of our compression approach must not exceeded the provided absolute precisions. A rough overview on the PPCA Offline method can be found in Figure 1.1.

In the situation where simulation results become available one after the other, the PPCA Online method can be applied, which is shown in Figure 1.2. For further information on the PPCA method, we refer to Chapter 4.

Since we do not exploit the topological relation in this method it is meaningful to apply a specified encoding scheme. This can be done by the induced Markov chain encoder. A short overview of its workflow can be found in Figure 1.3. In Chapter 5, the induced Markov chains are introduced in detail.

Finally, we combine the PPCA and iMc encoding, see Figure 1.4. The advantage of this combination is that during the optimization process the coding of graph-based data does not have to be carried out completely, but the expected size can be determined using entropy. This saves run time unlike when compression is fully executed.

1.2   Outline

In Chapter 2, we provide information about the content of crash test simulation results, how they are calculated and how they are managed by simulation data management systems (SDMS). Moreover, focus on the contents and properties of crash test simulation results.

In Chapter 3, we start with a short introduction to information theory and the resulting limitations of data compression. In addition, we depict the state-of-the-art compression methods in the context of simulation results. We distinguish between lossy and lossless compression and methods that exploit certain dependencies in time, space and as a new strategy similarities between sets of simulations. Moreover, we list the components of a state-of-the-art compression method in the context of crash test simulation results.

In Chapter 4, we introduce predictive principal component analysis (PPCA) that uses a linear dimensionality reduction method, namely PCA for the prediction of time-dependent data of simulation results. We distinguish between an offline method that is applied to all available data sets and an online method that can be used if the data sets are not available at the time of an initial compression. Since the PCA is a linear method and a crash test is a non-linear problem, we investigate how the residual of the prediction can be compressed efficiently.

Since we do not exploit the dependencies based on the topology of the finite element

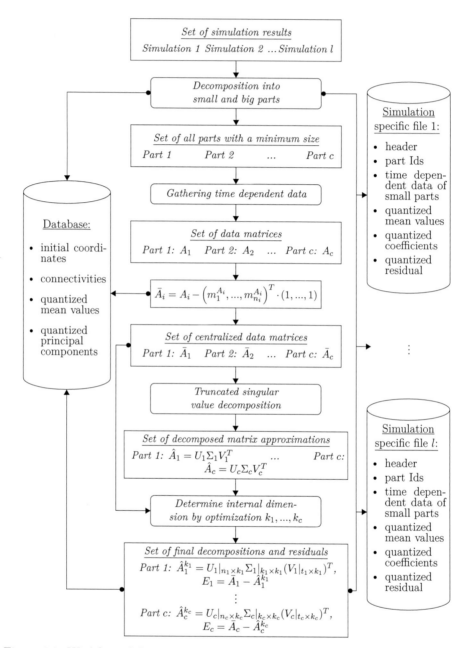

Figure 1.1: Workflow of the proposed compression method including the PPCA Offline method. We assume that two parts are identical if they are identical in the first time step.

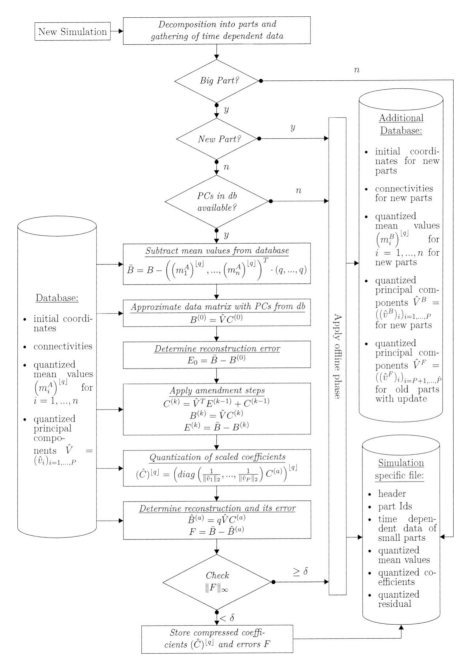

Figure 1.2: *Workflow of the proposed compression method including the PPCA Online method. In this example, we add one simulation result. In our example we have P number of principal components (PCs), the number of nodes n, and a amendment steps.*

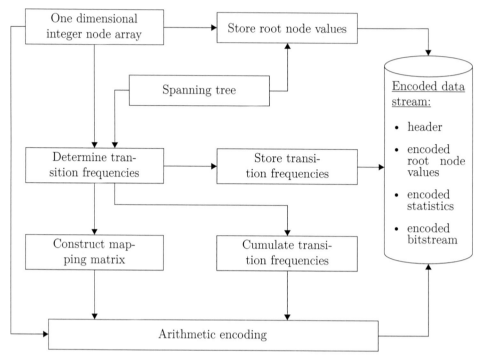

Figure 1.3: Diagram of the induced Markov chain workflow for one part.

mesh, we introduce the induced Markov chains (iMc) in Chapter 5. We begin with the application of the new induced Markov chains within an entropy encoder. Thereafter, we define the iMcs as a statistical model. We can combine the encoding scheme with different prediction methods that additionally reduce the dependencies in the investigated data sets. Moreover, we categorize and delimit the iMc from graphical models like Bayesian networks and Markov Random Fields (MRF).

In Chapter 6, we state our results for the PPCA for three benchmark data sets, see Section 6.2 and compare them with those of the state-of-the-art tool FEMZIP™ [124]. Furthermore, we combine the PPCA and the iMc and compare it with the combination of PPCA with zlib encoder [111] and the Rice encoder [110, 145].

In Chapter 7, we conclude with a short summary of our results and an outlook on future work.

1.3   Contributions

The most important part of this thesis is the development and elaboration of the mathematical background of the two procedures Predictive Principal Component Analysis (PPCA), see Section 4.3, and induced Markov chain (iMc) encoder, see Section 5.1, as

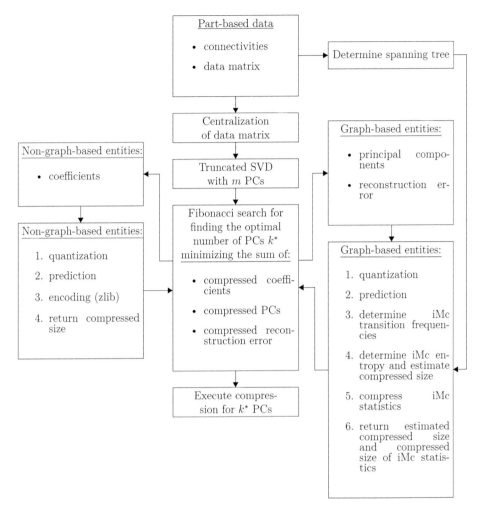

Figure 1.4: Combination of PPCA and iMc encoding. As part of the optimization process for graph based data, there is no need to actual encode the data, since iMc is an entropy encoder and the entropy can be determined applying the transition frequencies.

well as their combination, see Chapter 6. Both PPCA and the iMc encoder have already been published by the author, see [90, 96]. The PPCA is a machine learning technique that allows the reconstruction of compressed results with a given accuracy. This distinguishes the PPCA from other machine learning methods, which update a database on which decisions are made rather than extend it like the PPCA. An important contribution is made by the amendment steps, which make it possible to reuse the principal components meaningfully despite the lossy compression. A detailed differentiation of the PPCA from other procedures can be found in Section 4.1.

The iMc encoder is the first graph-based encoder that can also be used with a large alphabet. An important feature is that the overhead caused by the underlying model of the encoder can be estimated upwards. The estimation depends only on the topology of the graph, not on the alphabet. A differentiation of the iMc encoder from other methods can be found in Section 5.8.

The combination of PPCA offline method and iMc encoder proves to be the method that achieves the best compression rates. Due to the entropy encoder's property and the a priori determination of transition probabilities, the compressed size can be determined without having to perform the encoding completely. This is an advantage in the time-consuming optimization process determining the intrinsic dimension.

The contents of Chapter 2 and Chapter 3 are mainly known facts. Some evidence not available in the literature but important for this work is given, e. g. Theorem 3.42 and Theorem 3.63. Finally, we investigate the term "ergodicity" in the context of Markov chains and information theory since they are used in different ways, see Section 3.1.3.

1.4   Danksagung

Ich möchte mich an dieser Stelle bei all jenen bedanken, die mich bei der Anfertigung dieser Doktorarbeit unterstützt haben.

Ich danke Frau Prof. Dr. Caren Tischendorf für die fortwährende Betreuung dieser Arbeit, die Motivation den mathematischen Kern meiner Anwendung zu finden und diesen sauber zu definieren. Bedanken möchte ich mich bei Dr. Martin Weiser, der als Mentor im Rahmen der Berlin Mathematical School stets als Ansprechpartner bereit stand und der mit seinem Feedback und seiner kritischen Expertise diese Arbeit vorangebracht hat. Bei Prof. Dr. Rudolph Lorentz bedanke ich für einen Forschungsaufenthalt an der Texas A&M University at Qatar und seine engagierte Betreuung.

Mein ausdrücklicher Dank geht an Clemens-August Thole, der mich bestärkt hat PCA Verfahren für die Kompression von Scharen von Simulationsergebnissen zu untersuchen, wie auch seine fortwährende Unterstützung. Er hatte immer ein offenes Ohr für meine Fragen und Probleme.

Auf dem Weg zu dieser Arbeit haben mich viele ehemalige und aktuelle Kollegen unterstützt, kritisch hinterfragt und motiviert. Insbesondere sind hier Dr. Matthias Rettenmeier, Dr. Lennart Jansen und Stefan Mertler zu nennen bei denen ich mich ausdrücklich bedanken möchte. Auch danke ich Frau Yvonne Havertz für das detaillierte Lektorat.

Auch Danke ich der Berlin Mathematical School (BMS), die die Reisekosten meines Konferenzbesuchs auf der IEEE Konferenz ISSPIT in Noida Indien übernommen hat, sowie eine Plattform für viele spannende und gewinnbringende Bekanntschaften bietet. Neben meinen BMS Mentor, möchte ich Dr. Benjamin Trendelkamp-Schroer herausheben.

Besonders möchte ich mich bei meinen Eltern bedanken ohne deren immer während Unterstützung ich mein Studium in dieser Form nicht hätte durchführen können.

Letztlich gilt mein besonderer Dank meiner Frau Katrin und meinen Kindern Luisa und Mila, die es verstanden haben mich in allen Höhen und Tiefen meiner Arbeit aufzubauen, zu motivieren so dass ich diese Arbeit abzuschließen konnte.

Chapter 2

Crash test simulation results

This chapters gives a brief overview of crash test simulations, their content and of the features of the result files and their management.

A crash test simulation is the virtual imitation of a destructive vehicle crash test using numeric simulation. In the automotive industry, these simulations are central to Computer Aided Engineering (CAD). The first nonlinear finite element code, still in use today, for car crashes was initially published by Haug et al. in 1986 [59].

Since real crash tests are very time consuming and expensive the number of such tests that can be reasonably performed is limited. In the ideal case, real crash tests confirm the findings of previously performed simulations. The safety of new cars are tested in the region specific New Car Assessment Programme (NCAP), e.g. for Europe the European NCAP [41] with several types of crash tests being executed, see e.g. [41, 15]:

- frontal impact

- side impact

- rear impact

- rollover accidents

The NCAP real world tests are virtually modeled by discretizing the car with a finite element mesh. Modeling and meshing are time consuming tasks. Therefore, the modeling is done only once for the complete car and is used in all crash configurations. Figure 2.1 shows an example of a finite element mesh of a crash test. The simulation is followed by postprocessing on the results. The postprocessing, carried out by engineers, is partially automated, especially the determination of the crash relevant parameters and the output of automatically generated reports. Engineers also conduct manual postprocessing visualizing the simulation results in a specially developed software, the so-called postprocessor. The size of one simulation result file may range from one gigabyte (GB) to several GBs. Loading the data from a disk or transferring it via network can be a time consuming task. Therefore about 50% of all OEM (original equipment manufacturer) automotive companies apply the compression software FEMZIP [124, 132, 133] for the compression of the simulation result file. OEM car companies usually perform simulations in a standard workflow integrating compression immediately after simulation. The main advantages of postprocessing on compressed files are less disk space, bandwidth, and read-in times.

Figure 2.1: *Finite element mesh of a 2007 Chrysler Silverado crash test simulation [98]. The model contains 679 parts and 929,131 finite elements. "The model has been developed by The National Crash Analysis Center (NCAC) of The George Washington University under a contract with the FHWA and NHTSA of the US DOT".*

2.1 Calculation of crash test simulation results

In this section we provide a brief overview on the simulation of car crash tests. For more information regarding CAE we refer to [91], for solid mechanics to [147, 14], and for nonlinear finite element methods to [22, 144, 104]. A comprehensible introduction to simulation of crash tests and their history is given in [63]. Figure 2.2 gives an overview of the main components of crash test simulations.

Crash test simulations are part of solid mechanics. Two important properties of these simulations are that they deal with large displacements and forces so large that the threshold of linear elastic behavior of the material is exceeded [91, 73]. Therefore, the application of Hooke's law which states that "stress is proportional to the strain" is not reasonable for car crash simulations. In [63], Ibrahim concludes: "Simulation of vehicle accidents is one of the most challenging nonlinear problems in mechanical design as it includes all sources of nonlinearity." There are several factors that can cause nonlinearities to the physical process [104, 73, 91], namely

- material based nonlinearities

- geometric nonlinearities

- boundary based nonlinearities

2.1.1 Governing Equations

In the following section, we examine from which process the data that we will compress result. We investigate the governing equations of car crash simulation as defined in the LS-DYNA Theory manual [57]. We restrict our investigations to a simplified case with isotropic, linear elastic materials and small deformations. For a more general overview of this topic, we recommend [149] with a focus on the engineering field and Ciralet for the mathematical background.

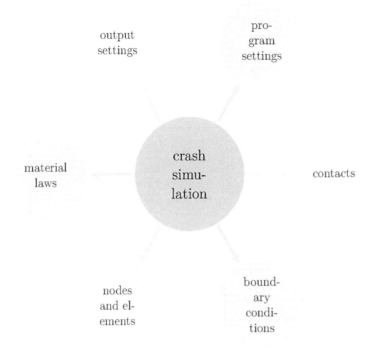

Figure 2.2: Central components of crash test simulation [91].

The field equations for solid mechanics are given by equilibrium (balance of momentum), strain-displacement relations, constitutive equations, boundary conditions, and initial conditions [147]. The linear momentum balance is a version of Newton's second law of motion, which states that the vector sum of the forces on an object is equal to the mass of that object multiplied by the object's acceleration vector. Forces working at boundaries, e g. fluid pressure, wind load or contact with other solids are called traction. Forces that work on the inside of solids e. g. gravity loading and electromagnetic forces are referred to as body forces.

Let $\mathcal{I} = [0, T]$ be a time interval and $\Omega \subset \mathbb{R}^3$ be the domain of a solid with a Lipschitz continuous surface Γ. Let $u : \Omega \times \mathcal{I} \to \mathbb{R}^3$ be the displacement function of a deformation. Moreover, let $\sigma : \Omega \times \mathcal{I} \to \mathbb{R}^{3 \times 3}$ be the Cauchy stress, $\rho : \Omega \times \mathcal{I} \to \mathbb{R}^3$ be the mass density, and $b : \Omega \times \mathcal{I} \to \mathbb{R}^3$ be the body force. The balance of linear momentum in terms of Cauchy stress is shown in Equation (2.1), see [14, 147, 148, 57],

$$\rho(\mathbf{x}, t)u_{tt}(\mathbf{x}, t) - \mathrm{div}\sigma(\mathbf{x}, t) - b(\mathbf{x}, t) = 0 \qquad (2.1)$$

At time $t = 0 \in \mathcal{I}$ and for $\mathbf{x} \in \Omega$ and $v_0 : \Omega \to \mathbb{R}^3$ we have the initial conditions

$$u(\mathbf{x}, 0) = \mathbf{x}$$
$$u_t(\mathbf{x}, 0) = v_0(\mathbf{x}).$$

For a meaningful definition of PDE (2.1), functions u, σ, and v_0 need to fulfill certain smoothness conditions.

Let $\Gamma^{(1)} \cup \Gamma^{(2)} = \Gamma$ with $\Gamma^{(1)} \cap \Gamma^{(2)} = \emptyset$ be the surface of Ω. We distinguish two types of boundary conditions. For surface $\Gamma^{(1)}$ we assume Dirichlet boundary conditions with a prescribed deformation $\bar{u}(\mathbf{x}, t) : \Gamma^{(1)} \times \mathcal{I} \to \mathbb{R}^3$. Moreover, we assume that a traction $\bar{t}(\mathbf{x}, t) : \Gamma^{(2)} \times \mathcal{I} \to \mathbb{R}^3$ is applied on surface $\Gamma^{(2)}$. Let $n = (n_1, n_2, n_3)^T \in \mathbb{R}^3$ denote the outer normal vector for a point $\mathbf{x} \in \Gamma^{(2)}$. The boundary conditions for PDE (2.1) are

$$u(\mathbf{x}, t) = \bar{u}(\mathbf{x}, t), \ \forall \mathbf{x} \in \Gamma^{(1)}, t \in \mathcal{I} \tag{2.2}$$

$$\bar{t}(\mathbf{x}, t) = \sigma(\mathbf{x}, t) n(\mathbf{x}, t), \ \forall \mathbf{x} \in \Gamma^{(2)}, t \in \mathcal{I} \tag{2.3}$$

In Figure 2.3, both the boundary conditions and the domains are marked. For isotropic,

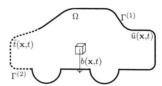

Figure 2.3: *Forces on a solid car with domain Ω. The body force $b(\mathbf{x}, t)$ acts on the inside of a solid. The union of $\Gamma^{(1)}$ and $\Gamma^{(2)}$ gives the surface of Ω. For surface $\Gamma^{(2)}$ traction forces are given, while for surface $\Gamma^{(1)}$ a prescribed deformation is known.*

linear elastic materials and small deformations, we have a direct relation between the Cauchy stress and the strain applying the 4th order stiffness matrix of material properties $C \in \mathbb{R}^{3 \times 3 \times 3 \times 3}$. This relation is known as Hooke's law [56]. We use the matrix notation from [148] that combines the first two and the last two indices as follows

$$s(1, 1) = 1, s(2, 2) = 2, s(3, 3) = 3, s(1, 2) = 4, s(2, 3) = 5, s(3, 1) = 6. \tag{2.4}$$

This notation is equivalent to the Voigt's notation, except for the sorting of the last three elements. We get $C_{ijkl} = \tilde{C}_{s(i,j)s(k,l)}$, $i, j, k, l \in \{1, 2, 3\}$, and a matrix $\tilde{C} \in \mathbb{R}^{6 \times 6}$. Let

$$\varepsilon_{ij}(\mathbf{x}, t) = \frac{1}{2} \left(\frac{\partial u_i(\mathbf{x}, t)}{\partial x_j} + \frac{\partial u_j(\mathbf{x}, t)}{\partial x_i} \right) \quad \text{with } i, j = 1, 2, 3 \tag{2.5}$$

be the components of vector $\varepsilon = (\varepsilon_{11}, \varepsilon_{22}, \varepsilon_{33}, 2\varepsilon_{12}, 2\varepsilon_{23}, 2\varepsilon_{31})^T$.
We get

$$\tilde{\sigma}(\mathbf{x}, t) = \sum_{j=1}^{6} \begin{pmatrix} \partial_1 u_1(\mathbf{x}, t) \tilde{C}_{1j} \\ \partial_2 u_2(\mathbf{x}, t) \tilde{C}_{2j} \\ \partial_3 u_3(\mathbf{x}, t) \tilde{C}_{3j} \\ (\partial_1 u_2(\mathbf{x}, t) + \partial_2 u_1(\mathbf{x}, t)) \tilde{C}_{4j} \\ (\partial_2 u_3(\mathbf{x}, t) + \partial_3 u_2(\mathbf{x}, t)) \tilde{C}_{5j} \\ (\partial_3 u_1(\mathbf{x}, t) + \partial_1 u_3(\mathbf{x}, t)) \tilde{C}_{6j} \end{pmatrix}. \tag{2.6}$$

To determine the deformation, we solve the partial differential equation (PDE) with respect to the displacement function $u(\mathbf{x}, t)$.
The momentum equilibrium, i.e. the balance of angular momentum yields the symmetry of Cauchy stress [14, 147]:

$$\sigma_{ij}(\mathbf{x}, t) = \sigma_{ji}(\mathbf{x}, t) \ \forall i, j = 1, 2, 3. \tag{2.7}$$

Let $\tilde{u}(\mathbf{x}) : \Omega \to \mathbb{R}^3$ with $\tilde{u} \in C_0^2(\Omega, \mathbb{R}^3)$ be a displacement field with $\tilde{u}(\mathbf{x}) = 0$ for all $\mathbf{x} \in \Gamma^{(1)}$. Let V and W be admissible spaces for $\tilde{u}$ and σ, respectively. We define a operator $P : W \to \tilde{W}$ with $P : \sigma(\mathbf{x}, t) \mapsto \tilde{\sigma}(\mathbf{x}, t)$. We further define the virtual strain with the help of virtual displacements as

$$(\tilde{\varepsilon}_{\tilde{u}})_{ij}(\mathbf{x}) = \frac{1}{2}\left(\frac{\partial \tilde{u}_i(\mathbf{x})}{\partial x_j} + \frac{\partial \tilde{u}_j(\mathbf{x})}{\partial x_i}\right), \ \forall i, j = 1, 2, 3 \tag{2.8}$$

2.1.1.1 Weak formulation

To find a solution for the PDE (2.1), we establish an integral formulation with the following property. If there is a solution of PDE (2.1), it is also a solution of the integral formulation. If we have a solution for the integral formulation, we interpret it as a weak solution of PDE (2.1).

Lemma 2.1 ([14]). *Let $u : \Omega \times \mathcal{I} \to \mathbb{R}^3$ be a solution of PDE (2.1). Then u also fulfills*

$$0 = \int_{\Omega} \tilde{u}^T(\mathbf{x})\rho(\mathbf{x}, t)u_{tt}(\mathbf{x}, t)d\Omega + \int_{\Omega} \tilde{\varepsilon}_{\tilde{u}}^T(\mathbf{x})\tilde{\sigma}(\mathbf{x}, t)d\Omega$$
$$- \int_{\Omega} \tilde{u}^T(\mathbf{x})b(\mathbf{x}, t)d\Omega - \int_{\Gamma^{(2)}} \tilde{u}^T(\mathbf{x})\bar{t}(\mathbf{x}, t)d\Gamma^{(2)} \tag{2.9}$$

for all displacement fields $\tilde{u} \in C_0^2(\Omega)$ with

$$\tilde{u}(\mathbf{x}) = 0, \ \forall x \in \Gamma^{(1)}. \tag{2.10}$$

Equation (2.9) is also called virtual work equation [14]. The integrations of Equation (2.9) are performed on the current geometry, see [57].

PROOF:

$$\int_{\Omega} \tilde{\varepsilon}^T(\mathbf{x})\tilde{\sigma}(\mathbf{x}, t)d\Omega$$

$$= \int_{\Omega} \tilde{\varepsilon}_{11}\sigma_{11} + \tilde{\varepsilon}_{22}\sigma_{22} + \tilde{\varepsilon}_{33}\sigma_{33} + 2\tilde{\varepsilon}_{12}\sigma_{12} + 2\tilde{\varepsilon}_{23}\sigma_{23} + 2\tilde{\varepsilon}_{31}\sigma_{31}d\Omega$$

$$\overset{(2.8)}{=} \int_{\Omega} \partial_1\tilde{u}_1\sigma_{11} + \partial_2\tilde{u}_2\sigma_{22} + \partial_3\tilde{u}_3\sigma_{33} + \partial_1\tilde{u}_2\sigma_{12} + \partial_2\tilde{u}_1\sigma_{12} + \partial_2\tilde{u}_3\sigma_{23}$$

$$+ \partial_3\tilde{u}_2\sigma_{23} + \partial_3\tilde{u}_1\sigma_{31} + \partial_1\tilde{u}_3\sigma_{31}d\Omega$$

$$\overset{\text{Gaußian,(2.7),(2.10)}}{=} \int_{\Gamma}\begin{pmatrix} \tilde{u}_1\sigma_{11} + \tilde{u}_2\sigma_{12} + \tilde{u}_3\sigma_{13} \\ \tilde{u}_1\sigma_{21} + \tilde{u}_2\sigma_{22} + \tilde{u}_3\sigma_{23} \\ \tilde{u}_1\sigma_{31} + \tilde{u}_2\sigma_{32} + \tilde{u}_3\sigma_{33} \end{pmatrix}^T nd\Gamma - \int_{\Omega} \tilde{u}_1\partial_1\sigma_{11} + \tilde{u}_2\partial_2\sigma_{22} + \tilde{u}_3\partial_3\sigma_{33}$$

$$+ \tilde{u}_2\partial_1\sigma_{12} + \tilde{u}_1\partial_2\sigma_{12} + \tilde{u}_3\partial_2\sigma_{23} + \tilde{u}_2\partial_3\sigma_{23} + \tilde{u}_1\partial_3\sigma_{31} + \tilde{u}_3\partial_1\sigma_{31}d\Omega$$

$$\overset{(2.3),(2.2)}{=} \int_{\Gamma^{(1)}} \tilde{u}_1 \sum_{j=1}^{3} \sigma_{j1} n_j + \tilde{u}_2 \sum_{j=1}^{3} \sigma_{j2} n_j + \tilde{u}_3 \sum_{j=1}^{3} \sigma_{j3} n_j d\Gamma^{(1)}$$

$$+ \int_{\Gamma^{(2)}} \tilde{u}_1 \sum_{j=1}^{3} \sigma_{j1} n_j + \tilde{u}_2 \sum_{j=1}^{3} \sigma_{j2} n_j + \tilde{u}_3 \sum_{j=1}^{3} \sigma_{j3} n_j d\Gamma^{(2)}$$

$$- \int_{\Omega} (\tilde{u}_1, \tilde{u}_2, \tilde{u}_3) \begin{pmatrix} \partial_1 \sigma_{11} + \partial_2 \sigma_{12} + \partial_3 \sigma_{13} \\ \partial_1 \sigma_{21} + \partial_2 \sigma_{22} + \partial_3 \sigma_{23} \\ \partial_1 \sigma_{31} + \partial_2 \sigma_{32} + \partial_3 \sigma_{33} \end{pmatrix} d\Omega$$

$$= 0 + \int_{\Gamma^{(2)}} \tilde{u}_1 \bar{t}_1 + \tilde{u}_2 \bar{t}_2 + \tilde{u}_3 \bar{t}_3 d\Gamma^{(2)} - \int_{\Omega} \tilde{u}^T \mathrm{div}\sigma d\Omega$$

$$= \int_{\Gamma^{(2)}} \tilde{u}^T \bar{t} d\Gamma^{(2)} - \int_{\Omega} \tilde{u}^T \mathrm{div}\sigma d\Omega \qquad (2.11)$$

We want to apply the Fundamental lemma of calculus of variations (FLCV), see e. g. [38]. We insert (2.11) in (2.9). For all $\tilde{u} \in C_0^\infty(\Omega)$ it holds:

$$\rho(\mathbf{x},t) u_{tt}(\mathbf{x},t) - \mathrm{div}\sigma(\mathbf{x},t) - b(\mathbf{x},t) = 0 \;\forall x \in \Omega, t \in \mathcal{I}. \qquad (2.12)$$

$$\Rightarrow \rho(\mathbf{x},t) u_{tt}(\mathbf{x},t) - \mathrm{div}\sigma(\mathbf{x},t) - b(\mathbf{x},t) = 0 \quad \text{almost everywhere in } \Omega$$

$$\Leftrightarrow \int_{\Omega} \tilde{u}^T(\mathbf{x})(\rho(\mathbf{x},t) u_{tt}(\mathbf{x},t) - \mathrm{div}\sigma(\mathbf{x},t) - b(\mathbf{x},t))d\Omega = 0$$

$$\Leftrightarrow \int_{\Omega} \tilde{u}^T(\mathbf{x})\rho(\mathbf{x},t) u_{tt}(\mathbf{x},t)d\Omega - \int_{\Omega} \tilde{u}^T(\mathbf{x})\mathrm{div}\sigma(\mathbf{x},t)d\Omega - \int_{\Omega} \tilde{u}^T(\mathbf{x})b(\mathbf{x},t)d\Omega = 0$$

$$\Leftrightarrow \int_{\Omega} \tilde{u}^T(\mathbf{x})\rho(\mathbf{x},t) u_{tt}(\mathbf{x},t)d\Omega + \int_{\Gamma^{(2)}} \tilde{u}^T(\mathbf{x})\bar{t}(\mathbf{x},t)d\Gamma^{(2)} - \int_{\Omega} \tilde{u}^T(\mathbf{x})\mathrm{div}\sigma(\mathbf{x},t)d\Omega$$

$$- \int_{\Omega} \tilde{u}^T(\mathbf{x})b(\mathbf{x},t)d\Omega - \int_{\Gamma^{(2)}} \tilde{u}^T(\mathbf{x})\bar{t}(\mathbf{x},t)d\Gamma^{(2)} = 0$$

$$\Leftrightarrow \int_{\Omega} \tilde{u}^T(\mathbf{x})\rho(\mathbf{x},t) u_{tt}(\mathbf{x},t)d\Omega + \int_{\Omega} \tilde{\varepsilon}_{\tilde{u}}^T(\mathbf{x})\tilde{\sigma}(\mathbf{x},t)d\Omega$$

$$- \int_{\Omega} \tilde{u}^T(\mathbf{x})b(\mathbf{x},t)d\Omega - \int_{\Gamma^{(2)}} \tilde{u}^T(\mathbf{x})\bar{t}(\mathbf{x},t)d\Gamma^{(2)} = 0$$

$\square$

We call solutions of (2.9) weak solutions and interpret them as an extension of solutions of PDE (2.2).
We define $\hat{\sigma} : V \to \tilde{W}$ with $\hat{\sigma} : u(\mathbf{x},t) \mapsto \tilde{\sigma}(\mathbf{x},t)$ and $\hat{\varepsilon} : C_0^\infty(\Omega) \to C_0^\infty(\Omega)$ with $\hat{\varepsilon} : \tilde{u}(\mathbf{x}) \mapsto \tilde{\varepsilon}(\mathbf{x},t)$. For all $t \in \mathcal{I}$ it holds

$$\int_{\Omega} \tilde{u}^T(\mathbf{x})\rho(\mathbf{x},t) u_{tt}(\mathbf{x},t)d\Omega + \int_{\Omega} \hat{\varepsilon}^T(\tilde{u}(\mathbf{x}))\hat{\sigma}(u(\mathbf{x},t))d\Omega$$

$$- \int_{\Omega} \tilde{u}^T(\mathbf{x})b(\mathbf{x},t)d\Omega - \int_{\Gamma^{(2)}} \tilde{u}^T(\mathbf{x})\bar{t}(\mathbf{x},t)d\Gamma^{(2)} = 0. \qquad (2.13)$$

Since the components of $\hat{\varepsilon}$ and $\hat{\sigma}$ are linear and combined in a linear way, $\hat{\varepsilon}$ and $\hat{\sigma}$ are also linear. Insertion gives the proof. We define

$$\mathcal{A}(u(\mathbf{x},t), \tilde{u}(\mathbf{x})) := \int_{\Omega} \hat{\varepsilon}(\tilde{u}(\mathbf{x}))\hat{\sigma}(u(\mathbf{x},t))d\Omega. \tag{2.14}$$

Since $\hat{\varepsilon}$ and $\hat{\sigma}$ are linear, $\mathcal{A}$ is a bilinear form.

2.1.1.2 Finite Element approach

Let $N \in \mathbb{N}$ be the number of nodes of the Finite Element Discretization with mesh points $\{\mathbf{x}^1, ..., \mathbf{x}^N\}$. Let $\phi_i \in V$ for $i = 1, ..., N$ be a nodal basis of a subspace $\hat{V} \subset V$. We choose an additional function $\phi_0 : \Omega \to \mathbb{R}^3$ that fulfills the inhomogeneous boundary condition (2.2). We apply the Galerkin method with

$$u(\mathbf{x},t) = \phi_0(\mathbf{x}) + \sum_{i=1}^{N} c_i(t)\phi_i(\mathbf{x}). \tag{2.15}$$

Since (2.9) holds for all $\tilde{u}(\mathbf{x}) \in C_0^{\infty}(\Omega)$ with $\tilde{u}(\mathbf{x}) = 0$ for all $\mathbf{x} \in \Gamma^{(1)}$, continuity of $\tilde{u}$, and C_0^{∞} being dense in H_0^1, it is also true for $\tilde{u}(\mathbf{x}) := \phi_j(\mathbf{x})$ for each $j = 1, ..., N$.

$$\mathcal{A}\left(\sum_{i=1}^{N} c_i(t)\phi_i(\mathbf{x}), \phi_j(\mathbf{x})\right) = \sum_{i=1}^{N} c_i(t)\mathcal{A}(\phi_i(\mathbf{x}), \phi_j(\mathbf{x}))$$
$$= Ac(t) \tag{2.16}$$

with $A = \mathcal{A}(\phi_j(\mathbf{x}), \phi_i(\mathbf{x}))_{ij=1,...,N}$ and $c(t) = (c_1(t), ..., c_N(t))^T$.
We get

$$\left\langle \rho(\mathbf{x},t)\sum_{i=1}^{N} c_i''(t)\phi_i(\mathbf{x}), \phi_j(\mathbf{x})\right\rangle = \sum_{i=1}^{N} c_i''(t)\langle\rho(\mathbf{x},t)\phi_i(\mathbf{x}), \phi_j(\mathbf{x})\rangle$$
$$= Mc''(t) \tag{2.17}$$

with

$$M(t) = ((\langle\phi_j(\mathbf{x}), \rho(\mathbf{x},t)\phi_i(\mathbf{x})\rangle)_{ij} \tag{2.18}$$

Moreover, we define

$$r^{(1)}(t) = (b_j(t))_j, \text{ with } b_j(t) = \langle f(t), \phi_j\rangle \wedge f(t)(\mathbf{x}) = b(\mathbf{x},t),$$
$$r^{(2)}(t) = (g_j(t))_j, \text{ with } g_j(t) = \int_{\Gamma^{(2)}} \phi_j^T(\mathbf{x})\bar{t}(\mathbf{x},t)d\Gamma^{(2)},$$
$$r(t) = r^{(1)}(t) + r^{(2)}(t). \tag{2.19}$$

With Equations (2.17), (2.16), and (2.19), we get the ODE

$$M(t)c''(t) + Ac(t) = r(t), \tag{2.20}$$

which is a linear inhomogeneous ordinary differential equation of second order. We rewrite Equation (2.20) as a system of ODEs of first order assuming that matrix M is regular. We start with the definition of the elements of the system:

$$y_1(t) := c(t)$$

$$y_2(t) := y_1'(t)$$
$$M(t)y_2(t)' := r(t) - Ay_1(t)$$

The system of ODEs of first order can be written as

$$\begin{pmatrix} I & 0 \\ 0 & M(t) \end{pmatrix} \begin{pmatrix} y_1 \\ y_2 \end{pmatrix}'(t) = \begin{pmatrix} y_2(t) \\ r(t) - Ay_1(t) \end{pmatrix} \qquad (2.21)$$

Usually, the inverse M^{-1} will not be calculated explicitly but a system of equations is solved with a linear solver.

In high dimensional crash simulations, the simulation time span is usually short. Therefore the crash simulation is usually solved with explicit methods. We use the explicit Euler [129] as an example:

$$\begin{pmatrix} I & 0 \\ 0 & M(t) \end{pmatrix} y(t) = \begin{pmatrix} y_1(t) \\ M(t)y_2(t) \end{pmatrix}, \quad f(y(t), t) = \begin{pmatrix} y_2(t) \\ r(t) - Ay_1(t) \end{pmatrix} \qquad (2.22)$$

Let the time increment $h = t_{n+1} - t_n$ be constant for all $n \in \mathbb{N}_0$ together with $t_n, t_{n+1} \in [0, T]$ for all $n \in \mathbb{N}_0$. Then the explicit Euler approximates the solution of the system of ODEs for time t_{n+1} by

$$\begin{pmatrix} y_1^{n+1} \\ M(t_n)y_2^{n+1} \end{pmatrix} = \begin{pmatrix} y_1^n \\ M(t_n)y_2^n \end{pmatrix} + h \begin{pmatrix} y_2^n \\ r(t_n) - Ay_1^n \end{pmatrix} \qquad (2.23)$$

The vector y_1^n approximates the coefficients of our Galerkin method:

$$y_1^n \approx c(t_n) \Leftrightarrow (y_1^n)_i \approx c_i(t_n) \ \forall i = 1, ..., N \qquad (2.24)$$

From (2.15) we know that for all mesh points $\mathbf{x}^j$ with $j \in \{1, ..., N\}$ as part of the finite element discretization and all $t_n \in [0, T]$ with $n \in \mathbb{N}_0$, we have

$$u(\mathbf{x}^j, t_n) = \phi_0(\mathbf{x}^j) + \sum_{i=1}^N c_i(t_n)\phi_i(\mathbf{x}^j) \stackrel{(2.24)}{\approx} \phi_0(\mathbf{x}^j) + \sum_{i=1}^N (y_1^n)_i\phi_i(\mathbf{x}^j).$$

As part of the solution process there is one value of the displacement function u for each mesh point $\mathbf{x}^j$ and each time step t_n available. Usually 60-80 time steps containing the data for all points of the finite element discretization are stored in a crash simulation result. The number of nodes is typically between 1 million and 10 million. For explicit methods, we need to guarantee the CFL condition.

Remark 2.2 ([63]). *"We can imagine a car crash as a wave propagation. In such cases explicit methods are reasonable since the CFL condition $\Delta t \leq \frac{l}{v_{acc}}$, where l is the element length and v_{acc} is the acoustic wave speed through the material of the element. E. g. for stainless steel, the acoustic speed is $v_{acc} = 5790\frac{m}{s}$ [39]. A common size of shell elements is around 5mm. Therefore, the maximal time step size is less than 1 microsecond."*

There are several finite elements to perform a spatial discretization of the partial differential equation (2.1). The main distinction lies in the dimension of the finite element. There are zero dimensional SPH elements, one dimensional beam, bar, and rod elements. For the 2D case there are shell and membrane elements and for three dimensions hexahedral and tetrahedral as well as less common pyramids and prisms. Figure 2.4 shows several two and three dimensional finite elements.

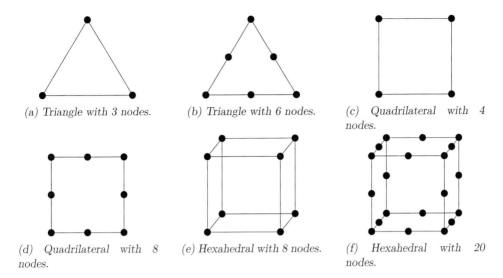

(a) Triangle with 3 nodes. (b) Triangle with 6 nodes. (c) Quadrilateral with 4 nodes.

(d) Quadrilateral with 8 nodes. (e) Hexahedral with 8 nodes. (f) Hexahedral with 20 nodes.

Figure 2.4: *Several two and three dimensional finite elements.*

2.1.2 Contact Algorithm

In this section, we give a brief overview on contact algorithms. They describe the usually nonlinear behavior of two contacting structural components. This is implicitly reflected by the equations in Section 2.1.1. The body forces contain results of the contact algorithm. For a deeper understanding of contact algorithms, we refer to the solid mechanics literature [147, 14] and the survey in [13].

The contact algorithm deals with the contact forces between structural parts. The differential equations of Section 2.1.1 are used to calculate the deformation of parts. However, since the engine's compartment space is limited, the components will hit against each other after some time. For safety reasons this is desired behavior, represented in numerics by contact algorithms.

Since there is a large number of contact algorithm methods [13], it is not possible to describe them in detail here. Therefore, we limit ourselves to the basic characteristics of contact algorithms and their effects on crash test simulations. In Figure 2.5, on the left we see an example of the so-called penalty method with the master mesh on the bottom and the slave mesh on the top. To determine the contact forces we need the distance

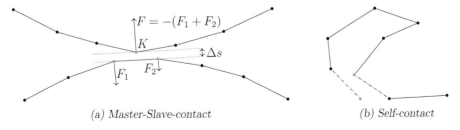

(a) Master-Slave-contact (b) Self-contact

Figure 2.5: *Two typical cases handled by a contact algorithm [91].*

from the slave nodes to the master segment. The force F acts on slave node K_1 and is dependent on the distance Δs and modeling parameters such as the contact density. On the master segment the force F will be distributed proportionally to its nodes with F_1 and F_2.

With big deformations or complex geometries components can have a self contact. In such cases, special types of algorithms are applied, which are usually computationally more expensive than the master-slave-approach [91]. This is one reason to model each part of a car individually and to avoid assemblies, since considerations about contacts of all nodes to every other node would be computationally too expensive. Therefore, only nodes with a distance smaller than a given threshold are considered. Especially in cases

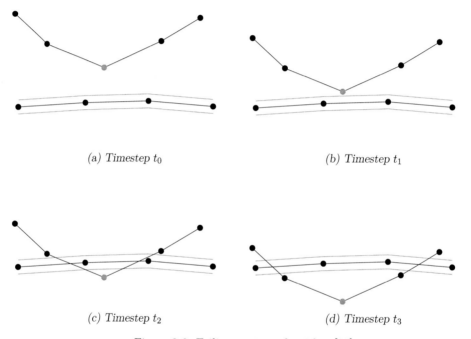

(a) Timestep t_0 (b) Timestep t_1

(c) Timestep t_2 (d) Timestep t_3

Figure 2.6: Failing contact algorithm [91].

of high speed, the contact algorithm can fail and thus two components penetrate. An example can be seen in Figure 2.6. The goal of having a functional contact algorithm and keeping the distance threshold small at the same time can only be achieved if the time step size is also small. Therefore explicit methods for solving the ODE are usually used for crash test simulations.

2.1.3   Best practice

As with all calculation models within the context of crash test simulations a best practice case can be determined. Usually, not everything is physically or mathematically justifiable but comes with more significant advantages such as less simulation time. The best practice presented here starts with the robustness of simulations and some remarks on the finite elements used. We will finish this section with some crash relevant parameters.

The properties as well as the network and boundary conditions are stored in a file, which is usually referred to as the Input Deck. The acceleration of the simulation results, also for commercial codes, can lead to two different results, even if you start with the identical input deck [91]. This is because in parallel calculations the results of different runs are summarized in the order different. Therefore, the deletion effects can be done in one run, while none are present in the other run. These small changes can uncover a bifurcation for an unstable case. A simple example for these types of bifurcations are Euler's buckling of columns [91], see Figure 2.7. The investigation of such behavior is called robustness

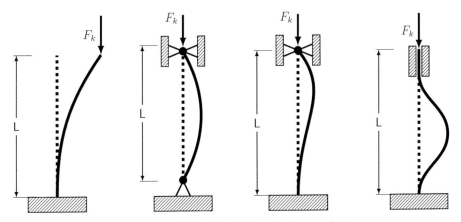

Figure 2.7: Euler's buckling of columns [91].

analysis [134]. It plays an important role in the development of safe cars because there are pre-defined tolerances in the production process. Different structural properties such as metal sheet thickness can give us some useful hints on the robustness of a design. For a robustness campaign several similar simulations have to be performed, a time-consuming task that produces a big bulk of similar simulation results [87].

As small effects like the order of summing interim results can have a big influence it is important to model the whole car with all of its components and at least in the crash area with a fine resolution. This includes components such as engine, battery etc. that usually are not deformed and for which so-called rigid body elements are used, which are generally more cost-efficient to simulate.

For the discretization with finite elements many different formulations of element types are applied, usually not reflected in the output format. Although 14 different finite element types for 2D shells exists [63, 57], in the result file all shell elements have the same description. A two dimensional discretization is less accurate than a three dimensional simulation of sheet metal parts, but significantly faster [91].

Relevant parameters such as the Head Injury Criterion (HIC) or the Tibia Index [41] are used to determine the crash worthiness of a car accident. The car design is optimized to fulfill the requirements of NCAP. To avoid foot and ankle injuries in frontal crashes the footwell intrusion is measured [41]. Therefore, relative and absolute distances are a critical parameter in the analysis of crash test simulation results.

Table 2.1: *Contents of a crash test simulation result in the d3plot format of LS-DYNA* [85].

Name	Description
Header	Control words
Initial geometry	Node coordinates
	Connectivities for 1D, 2D, and 3D elements
ID lists	References between sequential internal and user's numbering
State data	Coordinates, variables

Remark 2.3. *Lossy data compression can only be applied reasonably if the introduced loss, does not compromise the results of subsequent postprocessing.*

2.2   Contents and properties

This section collects properties of crash test simulation results that are important for applying lossy data compression. Some of these properties are analyzed in Section 2.1.1 and 2.1.3. In addition, we take a closer look at the different contents of a simulation result file to compress them successfully.

There are four widely used crash test simulation codes

- LS-DYNA [84]

- Pam-Crash [40]

- RADIOSS [5] and

- Abaqus [26].

All of them apply a finite element approach in combination with an explicit method in the time domain. The result file formats *DSY*, *Annn*, and *odb* of Pam-Crash, RADIOSS and Abaqus, respectively, are not to be published. Only the *d3plot* format of LS-DYNA is an open data format. Therefore we concentrate on LS-DYNA data sets. However, the three other result file formats are similar.

All output formats consist of the following data sets:

- header

- initial components, e.g. topology of the mesh, and

- time dependent data, e.g. coordinates, stresses.

Table 2.1 lists the contents of a crash test simulation result simulated with LS-DYNA and written in the *d3plot* format, see [85].

Normally, big car manufacturers store about 60 time steps per simulation. Therefore, the size of a simulation result is dominated by state dependent information such as coordinates and variables. The header and initial information such as ID lists and connectivities are time-independent and are stored only once in a simulation result file. Moreover, most of these data sets can efficiently be compressed by basic compression schemes. A successful compression of simulation results therefore depends on the compression of time-dependent data. While most of the common variables such as element failure can be compressed very efficiently, the compression of node coordinates is a challenge. The time-dependent data is usually defined on the elements of a mesh, e.g. nodes, edges, or finite elements.

Engineers usually keep the geometry of crashing cars for post processing, especially for visualization, and compare it to car crashes with design changes, e.g. in the context of robustness analysis, where the same finite element mesh description is a central requirement. If we want to increase the compression rate, we have to focus on the compression on the coordinates.

One crash test simulation can take from six hours to one day. With the increase of computational power the models can be refined to improve their accuracy. Since each simulation consumes a significant amount of time, computational power and energy, it is not a feasible option to simply reproduce the data by keeping the input deck. Therefore, simulation results are archived and stored as part of the development process. Moreover, certain simulation results must be stored for 10 years due to legal regulations.

A crash test model is usually composed of several thousand parts. This strategy enables:

- exchange of a component for different region specific designs, e.g. EU and US front spoiler.

- usage of many different materials with different material properties.

- usage of different sheet thicknesses (when modeled as shell) for each component.

- modeling of connections between different components e.g. by welding spots.

The majority of these parts consists of less than 50 nodes.

Moreover, some parts are auto generated by preprocessing tools. These parts can have a different connectivity description, even if at least in the first time step all node coordinates are in the same spot for different simulation runs. The topological description can be interpreted as a graph, which is important for the iMc encoder of Section 5.1.

Remark 2.4. *The main properties of crash test simulation results are:*

- *simulated data contains numerical errors.*

- *crash tests are a nonlinear task*

- *a model is divided in several hundred up to thousand parts.*

- *many of these parts have less than 40 nodes.*

- *some parts are auto generated.*

- *time-dependent data is stored on elements of the mesh.*

- *the topological description of the mesh must remain unchanged for postprocessing.*

- *the finite element mesh is dominated by 2D shell elements.*

- *in the finite element mesh a node is usually connected to less than 10 other nodes.*

In the following, we assume that all examined simulations have the properties stated in Remark 2.4.

2.3 Simulation data management systems

For several years SDMS have been applied successfully in the automotive industry. The advantages are obvious. Simulation data management systems help to avoid unnecessary simulations of the same model and make the mutual benefit from the work of colleagues. Some SDMS organize the simulations with the help of a development tree, see e.g. LoCo [118, 16]. This tree provides the information which parts were modified and which were not. If we have access to this information, we can decide, which results can be meaningfully compressed together. Moreover, SDMS has access to all simulation results that are started automatically. We can use these results and their similarities for compression. An important requirement for SDMS is the ability to apply deletion concepts for simulation results that are no longer required. If we integrate a compression method into an SDMS, we have to consider this condition as well. A simulation data management system usually

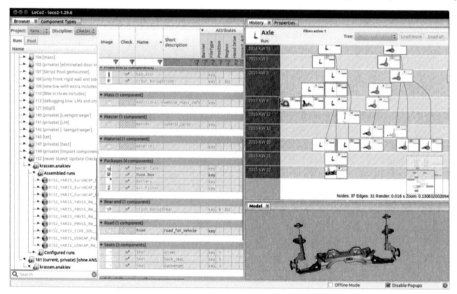

Figure 2.8: Development tree [16] represented by the software tool "LoCo 2" from SCALE GmbH [118].

has a server, where all simulated data is located and several clients that access the data remotely. Usually all engineers who share data work on a similar project and therefore the data is organized by projects. Moreover, all simulation results stored on a server are usually compressed. Engineers can access the data remotely or make a local copy of all necessary data sets on their local computer. When the data is compressed, they can save time when performing the data transfer.

Currently, there is no data compression software available, that exploits the similarities between simulation results and is plugged into a SDMS. Such a compression technique should have the properties, listed in Remark 2.5.

Remark 2.5. *A data compression method designed for a simulation data management system should have the following properties:*

- *The compression technique has to deal with the situation that the simulations come one after the other and are usually not immediately available.*

- *The compression technique should reuse information from previously compressed simulation results.*

- *The compression technique has to handle situations where we have modifications for a number of parts, but usually not for all parts.*

- *On the server the compression technique and the SDMS should maintain a database.*

- *The database files are not allowed to be updated, since this would require an update for all clients.*

- *A transfer of the database to the client should be beneficial for later data set transfers.*

- *It should be possible for us to synchronize the database with all clients at times with low network utilization.*

- *There should be the possibility to delete single simulation results.*

In the remainder of this thesis, we develop certain components to achieve a data compression workflow that perfectly fits into a SDMS.

Conclusion:

In this chapter, we state some well known facts of crash test simulation results. In Section 2.1, we briefly discuss the modeling of the crash and give an idea of the mathematical background. Here, everything is focused on obtaining a list of properties that is stated in Remark 2.4 and the main aspect of Section 2.2. In Section 2.3, we state some properties and ideas of simulation data management systems. Moreover, we draw a sketch of a data compression method that fits into a SDMS which is the basis for our PPCA method from Section 4.3.

Chapter 3

State-of-the-art compression methods for simulation results

In this chapter, we provide a short overview on data compression with a focus on compression of crash test simulation results and briefly explain its components. In addition we list the most relevant criteria to categorize these methods. We describe the Lorenzo predictor and tree differences as two well-known prediction methods. Moreover, we analyze entropy encoders and dictionary encoders as standard encoding schemes. First, we give a definition of data compression:

Definition 3.1 (data compression [116]). *Data compression is the process of converting an input stream, raw data, into another stream that has a smaller size.*

Next, we give the definition of compression ratio and compression factor.

Definition 3.2 (compression ratio, compression factor [116]). *The compression ratio is defined as*

$$compression\ ratio = \frac{size\ of\ the\ output\ stream}{size\ of\ the\ input\ stream}. \tag{3.1}$$

The compression factor is defined as

$$compression\ factor = \frac{size\ of\ the\ input\ stream}{size\ of\ the\ output\ stream}. \tag{3.2}$$

with the compression factor being the inverse of the compression ratio.

The entities compression factor and compression ratio are necessary to evaluate the performance of data compression.

In the following we want to provide a short overview of the application of lossy data compression to simulation results. A recent publication, see [52], provides a detailed overview of the applications of lossy data compression in the context of large PDE problems. In this publication, the application of compression to solve PDE problems, optimization using the Adjoint Equation, and the storage of result data for postprocessing are considered. In this thesis we focused on the third use case. Götschel gives in [52] three fields of application in which nowadays very large simulation results are reduced by lossy data compression:

1. Crash test simulations,

2. CFD (Computational Fluid Dynamics),

3. and weather and climate simulations.

In addition, in the fields of

1. NVH (Noise Vibration Harshness),

2. electromagnetic simulations

large amounts of data is generated nowadays.

In the field of crash simulations, there are the two commercial compression tools FEMZIP and SDMZIP. In the latter, the methods described in [90, 89, 88] and Chapter 4 of this thesis were implemented for productive applications. For CFD, there are two fee-based FEMZIP variants for the compression of simulation results in OpenFOAM and EnSight Gold format in addition to the approaches mentioned and developed by Götschel and Weiser. Cartesian grids are often used in the context of weather simulations, so that specialized tools like fpzip [83] or zfp [82] can be applied to these grids. In addition, Düben et al. shows in [33] two approaches for the compression of weather ensemble simulations. For compression, they suggest taking the ensemble mean or determining the precision used from the spread of the data. For NVH simulation results, there is also a FEMZIP variant compressing the common OP2 format. In the field of electromagnetic, the need for compression is not great despite large result files, since derived parameters, which make up only a small fraction of the initial quantity, are used for evaluation.

The structure of the next chapter is as follows. We, first, give on a short introduction to information theory with the introduction's focus on the best possible compression rates. We will see that the modeling of dependencies in the data is crucial for an efficient compression strategy. The reason for this is the entropy as the limit of lossless data compression, see Section 3.1. Subsequently, we provide the definitions necessary for a classification of data compression methods. We finish this chapter with an overview of data compression components that can be applied on simulation result files.

3.1 Information theory and its limits on data compression

In this section, we first provide a brief overview of information theory. For more information on this topic we refer to [143, 116, 25]. Unless specified else, definitions of Section 3.1 and 3.2 are taken from [116] also to be found in [143, 25, 93, 115]. In this context, we, additionally take some aspects from the viewpoint of probability theory into account. Its theoretical background can be found in Section A.1.1. Next we list the properties, and bounds on data compression. Finally, we model an input stream as a Markov chain.

The process of data compression cannot be applied unrestrictedly, since there are some fundamental limits on data compression which will be explained in the following. The main statement of this section is that a data stream can be compressed to almost its entropy. The entropy is therefore a measure of information content that is dependent on the underlying distribution of the data stream.

We start with the definition of an alphabet of a given stream.

Definition 3.3 (alphabet). *An alphabet $\mathcal{A}$ is the set of all possible symbols $x \in \mathcal{A}$ in a stream. All elements in $\mathcal{A}$ are pairwise distinct.*

For example, a stream can consist of the connectivities for one car component.

Example. For unsigned 32 bit integer, the complete alphabet would be

$$\mathcal{A} = \{0; 1; ...; 4, 294, 967, 295\},$$

where $2^{32} - 1 = 4, 294, 967, 295$.

In a stream of connectivities, we usually only encounter a small fraction of the used 32 bit integer alphabet. Therefore, occurring values could be represented by a shorter way with a new alphabet. This is the idea of encoding.

Definition 3.4 (encoding, codeword, decoding, code [107, 116]). *Let $\mathcal{A}$ and $\mathcal{B}$ denote two alphabets. An encoding c is a function*

$$c : \mathcal{A} \rightarrow \mathcal{B},$$

that assigns to all $a \in \mathcal{A}$ a codeword $c(a) \in \mathcal{B}$. If the mapping is injective, a unique inverse exists for the image $c(\mathcal{A})$:

$$c^{-1} : c(\mathcal{A}) \rightarrow \mathcal{A},$$

The function c^{-1} is called decoding. The entire set of codewords will be referred to as the code.

Assuming that both the encoder and the decoder operate sequentially, the previous example continuous as following.

Example. If all node ids of a connectivity stream are in the range of $\{0; 1; ...; 2^{16} - 1 = 65.535\}$ we can encode the stream by dismissing the leading 16 bits of each 32 bit integer. In this case, the alphabets are $\mathcal{A} = \{0; 1; ...; 4, 294, 967, 295\}$ and $\mathcal{B} = \{0; 1; ...; 2^{16} - 1\}$.

The example presented is simple and far from optimal as it only results in a data reduction of factor 2, due to the fact that, among others, every codeword has the same length. In the following, we will see that variable length codes can achieve distinctly better reductions, see also [116].

Let $(\Omega, \mathcal{B}(\Omega), \mathbb{P})$ be a probability space where $\mathcal{B}(\Omega)$ is the Borel-σ-algebra of Ω and $\mathbb{P}$ is a probability measure.

Definition 3.5 (random variable, events [74]). *Let $(\mathcal{A}, \mathcal{P}(\mathcal{A}))$ be a measurable space and $X : \Omega \rightarrow \mathcal{A}$ measurable.*

- *X is called random variable with values in $(\mathcal{A}, \mathcal{P}(\mathcal{A}))$.*

- *If $x \in \mathcal{P}(\mathcal{A})$ we will write $\{X \in x\} := X^{-1}(x)$ and $\mathbb{P}[X \in x] = \mathbb{P}[X^{-1}(x)] = \mathbb{P}_X[x]$.*

The sets $x \in \mathcal{P}(\mathcal{A})$ are called events.

We restrict our investigations to a finite alphabet $\mathcal{A}$.

Remark 3.6 ([119]). *The power set is the only σ-algebra which contains each elementary event $x = \{a\} \in \mathcal{P}(\mathcal{A})$ for $a \in \mathcal{A}$ for the finite set $\mathcal{A}$.*

Remark 3.6 and the restriction to a finite alphabet $\mathcal{A}$ justify the usage of the power set in Definition 3.5.

Definition 3.7 (distribution of a random variable [74]). *Let X be a random variable. The probability measure $\mathbb{P}_X := \mathbb{P} \circ X^{-1}$ is called distribution of X.*

We state the definition of the entropy for a random variable.

Definition 3.8 (entropy of a random variable [121, 143, 116]). *Let X be a discrete random variable with probability mass function $\mathbb{P}_X$ that takes values in a finite alphabet $\mathcal{A}$. The entropy of the random variable X is defined by*

$$H(X) = - \sum_{x \in \mathcal{A}} \mathbb{P}_X[x] \log_2(\mathbb{P}_X[x]) \tag{3.3}$$

In the case of $\mathbb{P}_X[x] = 0$ we set "$\mathbb{P}_X[x] \log_2(\mathbb{P}_X[x]) = 0$", which coincides with the limit of $\lim_{x \to 0} x \log_2(x)$.

Since we use logarithms to base 2 the entropy will be measured in bits. The entropy $H(X)$ quantifies the expected amount of information contained in a stream of randomly and hence independently appearing values from the finite alphabet $\mathcal{A}$.

Example. Let the alphabet be $\mathcal{A} = \{0, 1, 2, 3, 4, 5\}$ and the distribution of the random variable X be

$$\mathbb{P}_X[0] = \mathbb{P}_X[1] = \mathbb{P}_X[5] = \frac{1}{9}, \text{ and } \mathbb{P}_X[2] = \mathbb{P}_X[3] = \mathbb{P}_X[4] = \frac{2}{9}. \tag{3.4}$$

The entropy of the random variable X is

$$H(X) = - \sum_{x \in \mathcal{A}} \mathbb{P}_X[x] \log_2(\mathbb{P}_X[x])$$
$$\approx 2.503$$

The assumption that all values of an input stream can reasonably be modeled by independent and identically distributed (i.i.d., [see Definition A.11]) random variables cannot be applied here. Therefore, we investigate the more general stochastic processes, which is a family of random variables that are defined on the same probability space, see Definition A.12.

Definition 3.9 (joint entropy [25]). *The joint entropy $H(X, Y)$ of a pair of discrete random variables (X, Y) with measurable spaces $(\mathcal{A}, \mathcal{P}(\mathcal{A}))$ and $(\mathcal{B}, \mathcal{P}(\mathcal{B}))$ and a joint distribution $\mathbb{P}_{X,Y}$ is defined as*

$$H(X, Y) = - \sum_{x \in \mathcal{A}} \sum_{y \in \mathcal{B}} \mathbb{P}_{X,Y}[x, y] \log_2(\mathbb{P}_{X,Y}[x, y]). \tag{3.5}$$

In our investigations, the alphabets $\mathcal{A}$ and $\mathcal{B}$ are identical. Next, we define the entropy rate of a stochastic process.

Definition 3.10 (entropy rate [25, 37]). *The entropy rate of a stochastic process* $X = (X_t, t \in \mathbb{N}_0)$ *with events in* $\mathcal{P}(\mathcal{A})$ *is defined by*

$$H(X) = \lim_{n \to \infty} \frac{H(X_0, X_1, ..., X_{n-1})}{n}, \tag{3.6}$$

when the limit exists.

Working with the entropy rate is the chain rule for entropy is an indispensable tool. For that, we need the definition of conditional entropy, first.

Definition 3.11 (conditional entropy [54]). *The conditional entropy* $H(Y|X)$ *of a pair of discrete random variables* (X, Y) *with measurable spaces* $(\mathcal{A}, \mathcal{P}(\mathcal{A}))$ *and* $(\mathcal{B}, \mathcal{P}(\mathcal{B}))$ *and a joint distribution* $\mathbb{P}_{X,Y}$ *is defined as*

$$H(Y|X) = - \sum_{x \in \mathcal{A}} \sum_{y \in \mathcal{B}} \mathbb{P}_{X,Y}[X = x, Y = y] \log_2(\mathbb{P}_{Y|X}[Y = y|X = x]), \tag{3.7}$$

where $\mathbb{P}_{Y|X}[Y = y|X = x]$ *is the conditional probability of given that* x *happened and* y *will occur.*

Theorem 3.12 ([25]). *Let* X *be a random variable with alphabet* $\mathcal{A}$. *The chain rule for entropy is*

$$H(X_0, X_1) = H(X_0) + H(X_1|X_0) . \tag{3.8}$$

A short proof for the chain rule can be found in [25].

Remark 3.13 ([25]). *Let* $X = (X_t, t \in \mathbb{N}_0)$ *be i.i.d. random variables with* $X_t \in \mathcal{A}$. *This results in*

$$\begin{aligned} H(X) &= \lim_{n \to \infty} \frac{H(X_0, X_1, ..., X_{n-1})}{n} \\ &= \lim_{n \to \infty} \frac{n \cdot H(X_0)}{n} \\ &= H(X_0) \end{aligned}$$

The second equations hold due to the chain rule for the entropy. The entropy rate and the entropy coincide for i.i.d. random variables.

The relative entropy can be used to determine the overhead of encoding a stream that was generated by a distribution $\mathbb{P}_1$ but using a code that is optimal for distribution $\mathbb{P}_2$ [25].

Definition 3.14 (Relative entropy, Kullback-Leibler divergence [25]). *The relative entropy or Kullback-Leibler divergence between two probability mass functions* $\mathbb{P}_1$ *and* $\mathbb{P}_2$ *is defined as*

$$KL(\mathbb{P}_1, \mathbb{P}_2) = \sum_{x \in \mathcal{A}} \mathbb{P}_1[x] \log_2 \left(\frac{\mathbb{P}_1[x]}{\mathbb{P}_2[x]} \right) . \tag{3.9}$$

The Kullback-Leibler divergence with logarithms to base 2 can be interpreted as a measure for additional bits per symbol, which sums up with the entropy to the the number of bits that are needed for encoding.

We have to restrict ourselves to codes that are uniquely decodable, since we do not accept any distortion due to encoding. For fixed length codes this is given. With variable-length codewords we have to confine the set of all possible code books to the uniquely decodable cases. A sufficient prerequisite is to select prefix codes.

Definition 3.15 (prefix code [143]). *A code is called a prefix code or an instantaneous code if no codeword is a prefix of any other codeword.*

Remark 3.16 ([143]). *Prefix codes have the important property that they are uniquely decodable and easily identifiable in a decoder.*

To illustrate the meaning of prefix in the sense of codewords we state a code that does not fulfill the requirements for a prefix code.

Example. Let an alphabet $\mathcal{A} = \{0; 1; 2; 3; 4; 5\}$ be given. A code that does not satisfy the property of a prefix code can be found in Table 3.1.

Letter	Codeword
0	0000
1	1000
2	010
3	110
4	001
5	0100

Table 3.1: *Example for a code that is no prefix code. The codeword of symbol 2 is a prefix of the codeword of symbol 5.*

The codeword of symbol 2, i.e. 010, is a prefix of the codeword of symbol 5, which is **0100**.

As stated in [25]: "the class of uniquely decodable codes does not offer any further choices for the set of codeword lengths than the class of prefix codes. Hence, the bounds derived on the optimal codeword lengths continue to hold even when we expand the class of allowed codes to the class of all uniquely decodable codes."

Optimal codeword lengths are crucial for the success of data compression. To achieve a short code sequence, we have to ensure that the expected codeword lengths $l(x)$ for the elementary events $x = \{a\} \in \mathcal{P}(\mathcal{A})$, for $a \in \mathcal{A}$ are as small as possible [143].

3.1.1   Single random variables

In the following section, we state the relation between the entropy for a single random variable and the best possible compression ratio. The following theorem considers the compression of a single random variable. Applying the Kraft inequality [143, 25, 93], the code lengths for a binary representation with probability distribution $\mathbb{P}_X$ and values in a finite set $\mathcal{A}$ can be determined by:

$$l(x) = -\lceil \log_2(\mathbb{P}_X[x]) \rceil \tag{3.10}$$

Theorem 3.17 ([143]). *Let $X : \Omega \in \mathcal{A}$ be a random variable with probability distribution $\mathbb{P}_X$ and values in a finite set $\mathcal{A}$. Then a prefix code exists such that the minimum expected codeword length per symbol $L^* = \sum_{x \in \mathcal{A}} \mathbb{P}_X[x] l(x)$ of a random variable X satisfies*

$$H(X) \leq L^* < H(X) + 1 , \tag{3.11}$$

where equality occurs if and only if $l(x) = -\log_2(\mathbb{P}_X[x])$, i.e. the codeword lengths are equal to the ideal codeword lengths for all $x \in \mathcal{A}$.

Only if all probabilities are powers of two, the equality can occur .

Remark 3.18. *The expression $l(x) = -\log_2(\mathbb{P}_X[x])$ is also called optimal codeword length for the letter $x \in \mathcal{A}$ given the distribution $\mathbb{P}_X$.*

Example. Let $\mathbb{P}_X$ be the probability distribution for a random variable X with values in $\mathcal{A} = \{0; 1; 2; 3; 4; 5\}$ like in Equation (3.4). Having Remark 3.13 and Theorem 3.17 in mind and considering each element independently we get an expected code length per symbol in the range of $[2.503, 3.503)$. Using Equation (3.10) we can determine the code lengths for each elementary event and assign the codes with no codeword being the prefix of another codeword:

Letter/Event x	$\mathbb{P}_X[x]$	Code length $l(x)$ (3.10)	Codeword
0	$\frac{1}{9}$	4	0000
1	$\frac{1}{9}$	4	0001
2	$\frac{2}{9}$	3	010
3	$\frac{2}{9}$	3	011
4	$\frac{2}{9}$	3	100
5	$\frac{1}{9}$	4	0010

Table 3.2: *One possible prefix code for the distribution from (3.4) with code lengths determined with (3.10).*

The expected code length for the prefix code from Table 3.2 is

$$\begin{aligned} L^* &= E[l(x)] \\ &= \sum_{x \in \{0,\ldots,5\}} \mathbb{P}_X[x] l(x) \\ &= \frac{10}{3} \\ &= 3.\bar{3} \end{aligned}$$

Obviously, L^* fulfills the bounds given in Equation (3.11).
Interestingly, this is not the prefix code with the smallest expected code length. Using

a more advanced technique like the Huffman coding approach, see Section 3.3.3.1 and [93, 116, 115], results in e. g. the following code.

Letter/Event x	$\mathbb{P}_X[x]$	Code length $l_{\text{Huffman}}(x)$	Codeword
0	$\frac{1}{9}$	4	0000
1	$\frac{1}{9}$	4	0001
2	$\frac{2}{9}$	2	01
3	$\frac{2}{9}$	2	10
4	$\frac{2}{9}$	2	11
5	$\frac{1}{9}$	3	001

Table 3.3: *One possible prefix code for the distribution from (3.4) with code lengths determined by Huffman coding.*

The expected code length for the prefix code from Table 3.3 is

$$
\begin{aligned}
L^*_{\text{Huffman}} &= E[l_{\text{Huffman}}(x)] \\
&= \sum_{x \in \{0,\dots,5\}} \mathbb{P}_X[x] l_{\text{Huffman}}(x) \\
&= \frac{23}{9} \\
&= 2.\bar{5}
\end{aligned}
$$

Again, L^*_{Huffman} fulfills the bounds given in Equation (3.11) but is distinctly closer to the lower bound.

3.1.2 General stochastic processes

The stream, in general, shows high correlations between neighbored letters whereas the stream's behavior if its elements are not independent needs to be modeled as a general stochastic process.

Definition 3.19 (stationary [74]). *A stochastic process $X = (X_t)_{t \in \mathbb{N}_0}$ is called stationary if the distribution of random variables $X = (X_{t+s})_{t \in \mathbb{N}_0}$ is equal to $Y = (X_t)_{t \in \mathbb{N}_0}$ for all $s \in \mathbb{N}_0$.*

Theorem 3.20 ([25]). *Let $X = (X_t, t \in \mathbb{N}_0)$ be a stochastic process with $X_t \in \mathcal{A}$ and $\mathcal{A}$ finite. The minimum expected codeword length per symbol L^*_n of a stochastic process X satisfies*

$$
\frac{H(X_0, X_1, \dots, X_{n-1})}{n} \le L^*_n < \frac{H(X_0, X_1, \dots, X_{n-1})}{n} + \frac{1}{n}, \forall n \in \mathbb{N} \setminus \{0\} \tag{3.12}
$$

Moreover, if X is a stationary stochastic process, it holds that

$$
\lim_{n \to \infty} L^*_n = H(\mathcal{A}) , \tag{3.13}
$$

where $H(\mathcal{A})$ is the entropy rate of the model.

The proof is based on Kraft's inequality and the Asymptotic Equipartition Property (AEP), which is a direct consequence of the weak law of large numbers. The proof as well as more information to Kraft's inequality as well as the AEP can be found in [25]. Moreover, Theorem 3.20 is not only valid for i.i.d. sources but also for general stochastic processes [25].

Example. Let the probability distribution $\mathbb{P}_{X_t} = \mathbb{P}_X$ be for all random variables X_t, $t \in \{1; 2; ...; 10\}$) like in Equation (3.4). Let $X = (X_t, t \in \{1; 2; ...; 10\})$ be a stochastic process with $X_t \in \mathcal{A} = \{0; 1; 2; 3; 4; 5\}$. Let us assume, that all random variables are i.i.d. When we consider all 10 elements at once we get an expected length of the code in the range of $[2.503, 2.603)$, see Theorem 3.20:

$$\frac{H(X_1, X_2, ..., X_{10})}{n} \leq L_{10}^* < \frac{H(X_1, X_2, ..., X_{10})}{n} + \frac{1}{10}$$
$$\frac{10H(X_1)}{10} \leq L_{10}^* < \frac{10H(X_1)}{10} + \frac{1}{10}$$
$$2.503 \leq L_{10}^* < 2.603$$

Collecting 10 letters to one event, this will result in an alphabet size of $|\mathcal{A}|^{10}$. Theorem 3.17 and 3.20 clearly signal the importance of the entropy as a measure of information and a limit for data compression. With the original alphabet size being $|\mathcal{A}| = 6$ we would have to create for each of the $6^{10} = 60466176$ events one codeword, which is beyond the scope of this thesis. It, nonetheless, shows a way to enhance compression rates by improved modeling of the data.

The key to good encoding is to find and consider dependencies in the data. For streams, this will be done in Section 3.1.3 and for graph based data in Chapter 5.

3.1.3   Markov chains

One strategy to describe dependencies in a data stream is to model the stream as a Markov chain. This was initially done by Shannon in his famous paper [121] of 1948. This approach is both theoretically interesting and has successfully been applied in the compression algorithms Prediction by Partial Matching (PPM) and Dynamic Markov Coding (DMC) that will be specified in Section 3.3.3. Our new approach of modeling graph based data as a Markov chain, see Chapter 5, is also based on the following theory. An important condition for calculating a well defined entropy is, that the stochastic process, e.g. a Markov chain, is ergodic.

Remark 3.21. *In this section, we show that we need mean ergodicity for stochastic processes to reasonably investigate the entropy of an infinite long stream. This is due to the fact, that the AEP only holds in the mean ergodic case. Mean ergodicity is generally speaking the property that the overall averaged behavior is independent from the first events.*

Unfortunately, the concept of ergodicity is used differently in the field of Markov chains in contrast to information theory. The latter derives its definitions from the theory of dynamical systems. To prove this contention, we put a formal derivation of the ergodicity in the appendix, see Chapter B. In the following we derive ergodicity in the sense of Markov chains.

Definition 3.22 (Markov chain, Markov property [130]). *A Markov chain X on a countable state space $\mathcal{A}$ is a family of $\mathcal{A}$-valued random variables $(X_t, t \in \mathbb{N}_0)$ with the property that for all $t \in \mathbb{N}_0$ and $(i_0, ..., i_t, j) \in \mathcal{A}^{t+2}$, it holds that*

$$\mathbb{P}[X_{t+1} = j | X_0 = i_0, ..., X_t = i_t] = \mathbb{P}[X_{t+1} = j | X_t = i_t]. \tag{3.14}$$

Equation (3.14) is called Markov property.

Definition 3.23 (time homogeneous [130]). *A Markov chain X is called time homogeneous if and only if*

$$\mathbb{P}[X_{t+1} = j | X_t = i] = \mathbb{P}[X_1 = j | X_0 = i], \tag{3.15}$$

for all $t \in \mathbb{N}_0$ and all $i, j \in \mathcal{A}$.

Definition 3.24 (stochastic matrix [100]). *A matrix $P = (P_{ij})_{i,j \in \{1,...,m\}} \in [0,1]^{m \times m}$ is called stochastic if every row $(P_{ij})_{j \in \{1,...,m\}}$ is a distribution, i.e. $\sum_{j \in \{1,...,m\}} P_{ij} = 1$.*

In the following let X be a time homogeneous Markov chain with a finite state space $\mathcal{A} = \{i_1, ..., i_m\}$. Further let $P \in \mathbb{R}^{m \times m}$ be the matrix with transition probabilities from Definition 3.22 using the following notation:

Notation 3.25.
$$P_{ij} = P(j|i) = \mathbb{P}[X_1 = j | X_0 = i]. \tag{3.16}$$

Corrollary 3.26 ([130]). *An initial distribution ν together with a transition probability matrix P defines the distribution of a time homogeneous Markov chain.*

In the context of finite state space $\mathcal{A}$ and discrete time $t \in \mathbb{N}_0$, the demands on a stochastic process to be a Markov chain reduce to the Markov property that a transition probability matrix, which is a stochastic matrix, and a initial distribution ν on $\mathcal{A}$ exist.

Definition 3.27 (invariant distribution, stationary distribution [74]). *A distribution $\pi \in [0,1]^m$ on $\mathcal{A}$ is called invariant or stationary distribution in regard to a stochastic matrix P if*

$$\pi P = \pi. \tag{3.17}$$

Definition 3.28 (transient, recurrent, positive recurrent [130]). *Let a state $i \in \mathcal{A}$ be the starting state and T_i the first return time to i*

$$T_i = \inf\{t \in \mathbb{N} | X_t = i, X_0 = i\}. \tag{3.18}$$

Define further $\alpha_i^{(t)} = \mathbb{P}[T_i = t]$ as the probability that we return the first time to state i after t steps. Starting at state $i \in \mathcal{A}$ we call i transient if

$$\mathbb{P}[T_i < \infty] = \sum_{t=1}^{\infty} \alpha_i^{(t)} < 1. \tag{3.19}$$

The state $i \in \mathcal{A}$ is called recurrent if it is not transient. The mean recurrence time at state i is the expected return time

$$R_i = E[T_i] = \sum_{t=1}^{\infty} t\alpha_i^{(t)}. \tag{3.20}$$

The state $i \in \mathcal{A}$ is said to be positive recurrent if R_i is finite. A Markov chain is called positive recurrent if every state $i \in \mathcal{A}$ is positive recurrent.

A further property to categorize a Markov chain is the irreducibility. It is defined by an equivalence relation. Let be $i \in \mathcal{A}$ and $j \in \mathcal{A}$. If it holds $\mathbb{P}[X_t = j | X_0 = i] = \mathbb{P}_i[X_t = j] > 0$ for some $t \in \mathbb{N}\backslash\{0\}$, we use the notation $i \to j$ and call i leads to j.

Definition 3.29 (communicates [130])**.** *We say i communicates with j and write $i \leftrightarrow j$ if $i \to j$ and $j \to i$ hold.*

In [100], it is proved that $\leftrightarrow$ is an equivalence relation on a state space and thus partitions $\mathcal{A}$ into communicating classes.

Definition 3.30 (irreducibility [130])**.** *A Markov chain is said to be irreducible if $\mathcal{A}$ is a single communicating class regarding the equivalence relation $\leftrightarrow$.*

In an irreducible Markov chain it is possible to get from any state $i \in \mathcal{A}$ to any state $j \in \mathcal{A}$ of the state space [130].

Definition 3.31 (ergodicity of Markov chains [127])**.** *Let $\pi \in [0,1]^m$ be a invariant distribution with $\pi_i := \pi(\{i\}) > 0$ for all $i \in \mathcal{A}$. A Markov chain X is called ergodic if*

$$\lim_{t \to \infty} P^t(j|i) = \pi_j \ \forall i, j \in \mathcal{A}, \tag{3.21}$$

where $P^t(j|i) = \mathbb{P}[X_t = j | X_0 = i]$ describes the probability of being in state j after t-th time step starting in state i.

The notation $P^t(j|i)$ for $t \in \mathbb{N}$ and $i, j \in \mathcal{A}$ can be interpreted as the t-th power of the transition probability matrix P multiplied with the i-th unit vector e_i, since we start at state $i \in \mathcal{A}$ [60]. The right hand side of Equation (3.21) is independent of state $i \in \mathcal{A}$. Therefore, the probability to be in state j after a large number of time steps is independent of the initial value of X_0.

Lemma 3.32. *Let X be a positive recurrent irreducible Markov chain on the countable set $\mathcal{A}$ with an invariant distribution π. Then, $\pi_i := \pi(\{i\}) > 0$ for all $i \in \mathcal{A}$ and with $\mathbb{P}_\pi := \sum_{x \in \mathcal{A}} \pi_i \mathbb{P}_i$, it holds that X is stationary on $(\Omega, \mathcal{A}, \mathbb{P}_\pi)$ and ergodic.*

A short proof can be found in [74]. The probability π_i can be interpreted as the average duration of stays in $i \in \mathcal{A}$.

Definition 3.33 (mean ergodicity of Markov chains [55])**.** *A Markov chain X is called mean ergodic if*

$$\lim_{t \to \infty} \frac{1}{t} \sum_{k=0}^{t-1} P^k(j|i) = \pi_j \ \forall i, j \in \mathcal{A}. \tag{3.22}$$

Example. The following Markov chain is mean ergodic but not ergodic:

$$P = \begin{pmatrix} 0 & 1 \\ 1 & 0 \end{pmatrix} \text{ with } P^{2k-1} = P \text{ and } P^{2k} = I_2 \ \forall k \in \mathbb{N}$$

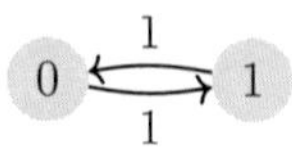

$$0 \underset{1}{\overset{1}{\rightleftarrows}} 1$$

Figure 3.1: *Graphical representation of a Markov chain with transition probabilities.*

Since we have two cluster points, the limit $\lim_{m \to \infty} P^m$ does not exist. Opposite to

$$\lim_{t \to \infty} \frac{1}{t} \sum_{k=0}^{t-1} P^k = \lim_{t \to \infty} \frac{1}{t} (\underbrace{I_2 + P}_{= \begin{pmatrix} 1 & 1 \\ 1 & 1 \end{pmatrix}} + \underbrace{I_2 + P}_{= \begin{pmatrix} 1 & 1 \\ 1 & 1 \end{pmatrix}} + \ldots + \underbrace{I_2 + P}_{= \begin{pmatrix} 1 & 1 \\ 1 & 1 \end{pmatrix}})$$

$$\underbrace{}_{= \frac{t}{2} \begin{pmatrix} 1 & 1 \\ 1 & 1 \end{pmatrix}}$$

$$= \frac{1}{2} \begin{pmatrix} 1 & 1 \\ 1 & 1 \end{pmatrix}$$

The invariant distribution of this example $\pi = (0.5, 0.5)$ can be easily calculated by Equation (3.17) and the requirement yields that π is a distribution, i.e. all elements are nonnegative and sum up to one.

Mean ergodicity is loosely speaking the average behavior over time and over the state space is identical with probability one.

Remark 3.34. *There is a difference in the meaning of ergodicity in the theory of Markov chains compared to the theory of dynamical systems. In Markov chains the term ergodicity coincides with the term mixing from the theory of dynamical systems. In dynamical systems the term ergodicity is used like the term mean ergodicity of the Markov chain theory, see Chapter B.*

In the following, we will list some necessary and sufficient properties for ergodicity.

Theorem 3.35 ([60]). *An irreducible Markov chain having a finite number of states is positive recurrent.*

Theorem 3.36 ([60]). *If a Markov chain with a finite state space $\mathcal{A}$ is irreducible, it will have a unique stationary distribution.*

Definition 3.37 (periodicity, aperiodicity [130]). *The period k of a state $i \in \mathcal{A}$ is defined as the greatest common divisor*

$$k = \gcd(\{n \in \mathbb{N} : \mathbb{P}[X_t = i | X_0 = i] > 0\}). \tag{3.23}$$

A state i is called aperiodic if k is equal to one. A Markov chain is called aperiodic if every state $i \in \mathcal{A}$ is aperiodic.

In the previous example the Markov chain is irreducible and the periodicity is 2.

Theorem 3.38 ([60]). *Let X be an irreducible positive recurrent Markov chain having an invariant distribution π.*

1. *If the chain is aperiodic, it holds that $\lim_{t\to\infty} P^t(j|i) = \pi_j \; \forall i,j \in \mathcal{A}$*

2. *If the chain is periodic with period d, then for all $i,j \in \mathcal{A}$ there exists an $r = r(i,j)$ with $0 \le r < d$ such that $P^t(j|i) = 0$ unless $t = kd + r$ for some $k \in \mathbb{N}$ and $\lim_{k\to\infty} P^{kd+r}(j|i) = d\pi_j \; \forall i,j \in \mathcal{A}$*

The first point in Theorem 3.38 states that an irreducible positive recurrent Markov chain has a unique stationary distribution. Applying the Theorem of Cauchy for a mean ergodic Markov chains we can show that the average converges against a unique stationary distribution [136]:

$$\lim_{m\to\infty} \frac{1}{m} \sum_{k=0}^{m-1} P^k(j|i) = \lim_{k\to\infty} \frac{1}{kd} \sum_{l=0}^{k-1} P^{ld+r}(j|i)$$

$$= \frac{1}{d} \lim_{k\to\infty} P^{kd+r}(j|i)$$

$$= \frac{1}{d} d \cdot \pi_j$$

Theorem 3.39. *An irreducible, positive recurrent and aperiodic Markov chain is ergodic.*

PROOF: This is obvious from Theorem 3.36, Theorem 3.38 and Definition 3.31. □

Theorem 3.40. *An irreducible, positive recurrent Markov chain is mean ergodic.*

PROOF: This is obvious from Theorem 3.36, Theorem 3.38 and Definition 3.33. □

To introduce our induced Markov chains in Chapter 5 we need the following statement.

Corrollary 3.41. *An irreducible Markov chain with finite state space is mean ergodic. If the Markov chain additionally fulfills aperiodicity, it is ergodic.*

PROOF: The first statement of the corollary follows directly by combining Theorem 3.35 and Theorem 3.40. The second one from combining 3.39 and Theorem 3.40. □

After the derivation of the ergodicity, we can now define the entropy of an ergodic Markov chain. We recall that the definition of the entropy rate is defined by Equation (3.6). In [25], Cover and Thomas stated that for an irreducible and aperiodic Markov chain X the entropy can be calculated by

$$H(X) = - \sum_{i,j\in\mathcal{A}} \pi_i P(j|i) \log_2 P(j|i) \ . \tag{3.24}$$

The Markov chain is ergodic, see Theorem 3.39. This requirement can be weakened to mean ergodic in the case of Markov chains.

Theorem 3.42. *The entropy for a time homogeneous mean ergodic Markov chain X with finite state space $\mathcal{A}$ can be calculated by the formula in (3.24).*

Theorem 3.42 was stated by Cover and Ekroot in 1993 [37], but without a formal proof. Interestingly, the requirement of aperiodicity was applied in the first edition of "Elements of information theory" by Cover and Joy (1991) and still can be found for the second edition of 2006 [25]. To the best of our knowledge, there is no proof available in literature. Even when the paper is cited the weakened requirements on the Markov chains are not exploited, e.g. [68]. To prove Theorem 3.42 first we state the chain rule for entropy from Theorem 3.12.

PROOF:

$$
\begin{aligned}
H(X) &= \lim_{n\to\infty} \frac{H(X_0, X_1, ..., X_{n-1})}{n} \\
&= \lim_{n\to\infty} \frac{1}{n} \sum_{k=1}^{n-1} H(X_k|X_{k-1}) + \frac{1}{n}H(X_0) \\
&= \lim_{n\to\infty} \frac{1}{n} \sum_{k=1}^{n-1} \sum_{i\in\mathcal{A}} \mathbb{P}[X_k = i] H(X_k|X_{k-1} = i) + \frac{1}{n}H(X_0) \\
&= \lim_{n\to\infty} \frac{1}{n} \sum_{k=1}^{n-1} \sum_{i\in\mathcal{A}} \mathbb{P}[X_k = i] H(X_1|X_0 = i) + \frac{1}{n}H(X_0) \\
&= \lim_{n\to\infty} \sum_{i\in\mathcal{A}} H(X_1|X_0 = i) \frac{1}{n} \sum_{k=1}^{n-1} \mathbb{P}[X_k = i] + \frac{1}{n}H(X_0) \\
&= \sum_{i\in\mathcal{A}} H(X_1|X_0 = i) \left(\lim_{n\to\infty} \frac{1}{n} \sum_{k=1}^{n-1} \mathbb{P}[X_k = i] \right) + \underbrace{\lim_{n\to\infty} \frac{1}{n}H(X_0)}_{=0}
\end{aligned}
$$

The second equation holds due to the chain rule for entropy and the Markov property. The fourth due to the time homogeneity. The fifth holds due to the finiteness of the state space $\mathcal{A}$. The sixth holds due to the existence of the limit of the summands. We have to consider the probability of $X_{k-1} = i$ for all $i \in \mathcal{A}$. Let ν be the initial distribution of the given Markov chain.

$$
\begin{aligned}
\mathbb{P}[X_k = i] &= \sum_{j\in\mathcal{A}} \mathbb{P}[X_0 = j, X_k = i] \\
&= \sum_{j\in\mathcal{A}} \mathbb{P}[X_0 = j]\mathbb{P}[X_k = i|X_0 = j] \\
&= \sum_{j\in\mathcal{A}} \nu(j)P^k(i|j)
\end{aligned}
$$

Since all probabilities $\mathbb{P}[X_k = i] \in [0,1]$ for all $k \in \mathbb{N}$ and $i \in \mathcal{A}$, we have a finite sum $\frac{1}{n}\sum_{k=1}^{n-1} \mathbb{P}[X_k = i]$. Therefore, we are allowed to exchange the order of summation as in the following.

$$
\lim_{n\to\infty} \frac{1}{n} \sum_{k=1}^{n-1} \mathbb{P}[X_k = i] = \lim_{n\to\infty} \frac{1}{n} \sum_{k=1}^{n-1} \sum_{j\in\mathcal{A}} \nu(j)P^k(i|j)
$$

$$= \sum_{j \in \mathcal{A}} \nu(j) \lim_{n \to \infty} \frac{1}{n} \sum_{k=1}^{n-1} P^k(i|j)$$

$$= \sum_{j \in \mathcal{A}} \nu(j) \pi_i$$

$$= \pi_i$$

Due to ν is a distribution and the Markov chain is mean ergodic. Hence,

$$H(X) = \sum_{i \in \mathcal{A}} H(X_1|X_0 = i) \left(\lim_{n \to \infty} \frac{1}{n} \sum_{k=1}^{n-1} \mathbb{P}[X_k = i] \right) + 0$$

$$= \sum_{i \in \mathcal{A}} H(X_1|X_0 = i) \pi_i$$

$$= \sum_{i \in \mathcal{A}} \pi_i \left(- \sum_{j \in \mathcal{A}} \mathbb{P}[X_1 = j|X_0 = i] \log_2(\mathbb{P}[X_1 = j|X_0 = i]) \right)$$

$$= - \sum_{i,j \in \mathcal{A}} \pi_i \mathbb{P}[X_1 = j|X_0 = i] \log_2(\mathbb{P}[X_1 = j|X_0 = i])$$

$\square$

Based on Theorem 3.42 we obtain the following Corollary.

Corollary 3.43. *For a well defined entropy of a Markov chain X, it is sufficient to have the Markov chain property ergodicity with its definition from dynamical systems. The Markov property aperiodicity is not necessary. Compared to the definition of ergodicity in Markov chains, we can neglect the aperiodicity.*

The next example demonstrates that the periodicity of an irreducible Markov chain is no issue for the calculation of the entropy.

Example. Let a Markov chain with the following graphical representation be given:

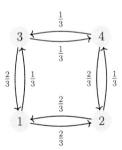

Figure 3.2: Graphical representation of a Markov chain with transition probabilities.

The transition probabilities can be expressed by matrix P.

$$P = \begin{pmatrix} 0 & \frac{2}{3} & \frac{1}{3} & 0 \\ \frac{2}{3} & 0 & 0 & \frac{1}{3} \\ \frac{2}{3} & 0 & 0 & \frac{1}{3} \\ 0 & \frac{2}{3} & \frac{1}{3} & 0 \end{pmatrix} \quad \text{with } P^{2k} = \begin{pmatrix} \frac{2}{3} & 0 & 0 & \frac{1}{3} \\ 0 & \frac{2}{3} & \frac{1}{3} & 0 \\ 0 & \frac{2}{3} & \frac{1}{3} & 0 \\ \frac{2}{3} & 0 & 0 & \frac{1}{3} \end{pmatrix} \quad \text{and } P^{2k-1} = P \ \forall k \in \mathbb{N} \backslash \{0\}$$

The unique invariant distribution of this Markov chain is $\pi = (1/3, 1/3, 1/6, 1/6)^T$.
We assume an initial distribution of $\nu = (0.4, 0.1, 0.1, 0.4)$. This results in two "quasi-invariant" distributions

$$\lim_{k \to \infty} \nu P^{2k} = \pi_1^\nu = \left(\frac{8}{15}, \frac{2}{15}, \frac{1}{15}, \frac{4}{15} \right)$$

$$\lim_{k \to \infty} \nu P^{2k-1} = \pi_2^\nu = \left(\frac{2}{15}, \frac{8}{15}, \frac{4}{15}, \frac{1}{15} \right)$$

for all $k \in \mathbb{N}$.
For the arithmetic mean of the two quasi-stationary distributions, it holds:

$$\frac{\pi_1^\nu + \pi_2^\nu}{2} = \left(\frac{1}{3}, \frac{1}{3}, \frac{1}{6}, \frac{1}{6} \right)^T = \pi \ . \tag{3.25}$$

The entropy can be calculated with the formula (3.24):

$$H(X) = - \sum_{i,j=1}^{4} \pi_i P_{ij} \log_2 P_{ij}$$

$$= -\frac{2}{3} \log_2 \frac{2}{3} - \frac{1}{3} \log_2 \frac{1}{3}$$

$$\approx 0.918.$$

In each step of the Markov chain, there are only two possible next states. This can be exploited in an encoding of a stream that is produced by the Markov chain X. If we start e.g. with $X_0 = 1$ the next state is with probability $2/3$ event $X_1 = 2$ and with probability $1/3$ event $X_1 = 3$. If we apply an adaptive encoder that switches the applied distribution in each state we can almost achieve the entropy rate as bits per state after encoding, see Section 3.3.3.

The question arises if we can also neglect the irreducibility. The answer is generally no as demonstrated in the following example:

Example. Let a Markov chain with the following graphical representation be given: The transition probabilities can be expressed by the matrix P.

$$P = \begin{pmatrix} P_1 & 0 \\ 0 & P_2 \end{pmatrix} \quad \text{with } P_1 = \begin{pmatrix} 0 & \frac{1}{3} & \frac{2}{3} \\ \frac{5}{6} & 0 & \frac{1}{6} \\ \frac{1}{2} & \frac{1}{2} & 0 \end{pmatrix} \quad \text{and } P_2 = \begin{pmatrix} 0 & \frac{1}{2} & \frac{1}{2} \\ \frac{1}{2} & 0 & \frac{1}{2} \\ \frac{1}{2} & \frac{1}{2} & 0 \end{pmatrix}$$

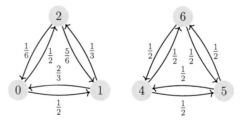

Figure 3.3: *Graphical representation of a Markov chain with transition probabilities.*

In this case, the Markov chain is not irreducible and there is no unique invariant distribution. All possible invariant distributions lie in the span of

$$\pi^{(1)} = \left(\frac{33}{83}, \frac{24}{83}, \frac{26}{83}, 0, 0, 0\right)^T \text{ and } \pi^{(2)} = \left(0, 0, 0, \frac{1}{3}, \frac{1}{3}, \frac{1}{3}\right)^T ,$$

with nonnegative components that sum up to one. We can now calculate the entropy for the invariant distributions $\pi^{(1)}$ and $\pi^{(2)}$.

In the first case let the initial distribution deterministically start in the irreducible subgraph X_1 represented by P_1:

$$
\begin{aligned}
H(\mathcal{A}_{\pi^{(1)}}) &= -\sum_{i,j=1}^{6} \pi_i^{(1)} P_{ij} \log_2 P_{ij} \\
&= -\frac{33}{83}\left(\frac{1}{3}\log_2\frac{1}{3} + \frac{2}{3}\log_2\frac{2}{3}\right) - \frac{24}{83}\left(\frac{5}{6}\log_2\frac{5}{6} + \frac{1}{6}\log_2\frac{1}{6}\right) - \frac{26}{83}\left(2\frac{1}{2}\log_2\frac{1}{2}\right) \\
&\approx 0.866.
\end{aligned}
$$

In contrast to the entropy for deterministically starting in 1, 2 or 3 the entropy for starting in 4, 5 or 6 is:

$$
\begin{aligned}
H(\mathcal{A}_{\pi^{(2)}}) &= -\sum_{i,j=1}^{6} \pi_i^{(2)} P_{ij} \log_2 P_{ij} \\
&= -\frac{1}{3}\left(\frac{1}{2}\log_2\frac{1}{2} + \frac{1}{2}\log_2\frac{1}{2}\right) - \frac{1}{3}\left(\frac{1}{2}\log_2\frac{1}{2} + \frac{1}{2}\log_2\frac{1}{2}\right) - \frac{1}{3}\left(\frac{1}{2}\log_2\frac{1}{2} + \frac{1}{2}\log_2\frac{1}{2}\right) \\
&= 1.
\end{aligned}
$$

If it is uncertain which connected component we start with, the entropy will be between $H(\mathcal{A}_{\pi^{(1)}})$ and $H(\mathcal{A}_{\pi^{(2)}})$.

The previous example shows that the initial distribution influences the entropy in a non mean ergodic cases. Therefore, the entropy is not well defined in such cases and will not give us a bound on the best possible compression ratios.

Remark 3.44. *The entropy gives us a quantification about the mean information content of a Markov chain. As the periodicity of all states is always the same in an irreducible Markov chain, we can consider the subsequences for every $0, ..., d-1$ for its own if d is the periodicity.*

Remark 3.45 ([25]). *If a Markov chain is irreducible and aperiodic, it has a unique stationary distribution on the states and any initial distribution tends to the stationary distribution as $n \to \infty$. In this case, even though the initial distribution is not the stationary distribution, the entropy rate, defined in terms of long-term behavior, can be determined by Formula (3.24).*

We can generalize Remark 3.45 in the case of periodicity of the Markov chain. From Theorem 3.36, we know that aperiodicity of a Markov chain is not necessary for uniqueness of the stationary distribution. Moreover, considering that periodicity $d \in \mathbb{N}\backslash\{0, 1\}$ of all states in an irreducible Markov chain is identical, see Remark 3.44, we can split all states into subsequence $X^{(1)}, X^{(2)}, ..., X^{(d)}$, i.e. Submarkov chains. When we apply Remark 3.45 to all Submarkov chains and assume an initial distribution such that it is valid for the considered Submarkov chain, we get the convergence to the stationary distribution $\pi^{(i)}$ for each Submarkov chain $X^{(i)}$, $i = 1, ..., n$. The unique stationary distribution of the overall Markov chain π is then given by:

$$\pi = \frac{1}{d} \sum_{i=1}^{d} \pi^{(i)} \tag{3.26}$$

Remark 3.46. *In the case of an irreducible but not aperiodic Markov chain, the convergence speed of an arbitrary initial distribution is slower by the d-th root compared to the aperiodic case, since each of the Submarkov chains has to converge for its own.*

Therefore, only mean ergodicity is necessary to have a well defined entropy of a Markov chain, but the long term behavior is reached earlier in case of ergodicity. The entropy will be calculated based on the long term behavior that usually does not coincide with the initial distribution, For our practical implementation fast convergence is desirable as we develop an approach based on the entropy and our encoder has to also work properly for short messages.

3.2   Types of data compression

There are various categories of data compression that can be used for classification. In this section, we will state well known criteria important to categorize our PPCA method in Section 4.3 as well as our iMc encoder in Section 5.1. Unless specified differently, basic definitions, nomenclature, and categorizations of Section 3.2 are taken from [116] but can also be found in [143, 25, 93, 115].
The actual compression will be performed by a so-called encoder.

Definition 3.47 (encoder, decoder [116]). *An encoder is a program that compresses raw data of an input stream and creates an output stream with compressed data. A decoder is a program that decompresses a stream with compressed data.*

A famous encoder is for example the Huffman encoder of Section 3.3.3.1, which can also be combined with various preprocessing steps like prediction methods of Section 3.3.2. A very important distinguishing criterion for data compression methods is whether the compression is executed lossy or losslessly.

Definition 3.48 (lossy, lossless compression [116]). *With a lossless data compression the input stream can be exactly reproduced while decompression. For a lossy compression, the data is only up to a certain precision reproducible.*

In Section 3.3.1, we introduced the quantization that is a typical preprocessing step of a lossy compression scheme. Wherever a loss of information is permissible, a lossy data compression is almost always meaningful, since the best possible compression rates are better than in the lossless case, see rate-distortion theory in [25, 53].
A further property for categorization is, if the compression method adapts to a given data set.

Definition 3.49 (adaptive, nonadaptive compression method [116]). *A nonadaptive compression method is called rigid and does not modify its operations, parameters, or tables in response to the particular data being compressed. In contrast, an adaptive method examines the raw data and modifies its operations and parameters accordingly.*

Generally, adaptive compression schemes gain better compression rates at the cost of higher run times [93]. In Sections 3.3.3.3 and 3.3.3.5, two famous representatives of adaptive encoders are described, namely prediction by partial matching (PPM) and dynamic Markov coding (DMC).

A further category for compression techniques is whether the computational complexity of compression and decompression are similar.

Definition 3.50 (symmetrical and asymmetrical compression [116]). *We speak of symmetrical compression if the compressor and decompressor use the same algorithm for encoding and decoding of the data. An important property is that the complexity of compression and decompression is the same. In contrast asymmetric compressions differ in time complexity.*

An example for an asymmetric encoder is LZ77 [150], whose decoder is much simpler than the encoder [116]. In the context of compression of simulation results, the compression phase may last longer than the decompression phase, see Remark 2.4.

Furthermore, encoders can be categorized if they use the knowledge of a data type.

Definition 3.51 (physical and logical compression method [116]). *A physical compression method only investigates the input stream and tries to find reoccurring bit stream samples. These samples will get a short representation in the encoded stream. The encoder ignores the meaning of data items from the input stream. In contrast, logical methods look in the stream for frequently upcoming items and replace them with shorter codes.*

A typical physical compression method is the LZ77 method, which interprets each in input data as a plain bit stream. In opposite to LZ77, the Rice encoder [145] assumes unsigned 32 Bit integer data and is a therefore a logical compression method.

Compression methods are categorized by being lossy or lossless. A lossy compression achieves generally higher compression factors [116]. Therefore, it is applied if a 100% reproduction of the input stream is not necessary, e.g. for the numerically calculated time dependent data of simulation results. For data sets such as connectivities of a finite element mesh engineers usually do not accept any deviation. Hence, we have to apply a lossless compression method.

Next, we define four different type of encoders and categorize them afterwards.

Definition 3.52 (Run-length encoding [116]). *Let $\mathcal{A}$ be a finite alphabet. Run-length encoders scan the input stream for consecutive reoccurring data items. If an item $d \in \mathcal{A}$ occurs $n \in \mathbb{N}\backslash\{0\}$ consecutive times in the input stream, replace the n occurrences with the single pair "nd". The n consecutive occurrences of a data item are called run-length of n. This method assigns fixed length codes to the symbols it operates on.*

The functionality of run-length encoding is e.g. available in the widely used gzip compression tool [44].

Definition 3.53 (Statistical encoder, entropy encoder [116]). *Statistical encoders determine or assume a distribution for elements of the input stream. Entropy encoders achieve an almost optimal compression rate bound by the entropy. Based on a given or determined distribution an entropy encoder is applied to achieve the actual compression.*

Well-known statistical encoders are Huffman encoders and arithmetic encoders, see Section 3.3.3.1.

We distinguish between methods, where we apply a previously fixed distribution, e. g. Golomb encoder [47], and methods where a data dependent distribution is determined. In the latter allows to find among the various distributions the fits the optimal distributions. For both approaches, there is no need to store the distribution as it is part of the encoder and decoder, respectively. Furthermore, we can calculate a distribution based on the data with the need of storing the statistics as well as the encoded data.

Definition 3.54 (Dictionary-based encoding [116]). *Dictionary-based encoders select strings of symbols and encode each string as a token using a dictionary. The dictionary holds strings of symbols and it may be static or dynamic.*

Famous dictionary based encoders are the LZ77 [150] and the Deflate encoder [116]. The latter is part of gzip [44] and zlib [111] compression tools.

Definition 3.55 (Transform encoding [116]). *A transform encoder applies a transformation and encodes the transformed values instead of the original ones.*

Famous representatives of transform encoders are discrete cosine transformation (DCT) [1] and wavelet coding [31, 30].

3.3 Components of data compression

In this section, we describe the structure of a data compression workflow. Wherever it is meaningful, we apply a lossy data compression, since the achieved compression rates are better than in the lossless case, cf. rate-distortion theory [25, 53].

The components of a state-of-the-art lossy compression method in the context of crash test simulation results are, see [132, 133]:

1. quantization,

2. prediction or transformation, and

3. encoding.

In case of a transformation, the first and the second point can be switched, see [141, 49, 125]. For the compression of simulation results we distinguish between data sets that have to be reconstructed without any error and data sets that can be reconstructed up to a certain precision. For car crash simulations for example, the topological description of the mesh has to be restored one to one. Therefore, we perform a lossless data compression for the mesh topology. Since the topology is described by integer ids we will not perform a quantization, which distinguishes the compression workflow from those of numerically calculated time dependent data, e.g. coordinates or stresses, see Remark 2.4, which generates most of the data in a crash test simulation result file.

The main components of data compression for crash tests are briefly described in the following.

3.3.1 Quantization

The quantization is a process to restrict a variable quantity to discrete values rather than to a continuous set of values [116]. In the case of compressing floating point values, the quantization maps the set of all representable floating point numbers on a computer to the set of integers. In general, this process is lossy and generates an irreversible error. As time dependent data of simulation results is calculated numerically it is only precise up to a certain accuracy, see Remark 2.4. This justifies a lossy compression up to a respective absolute precision. In practice, the accuracy needed is estimated by expert users, e. g. for post processing.

Let $Q \in \mathbb{R}$ be given. The quantization is a function defined as

$$\text{quant}_Q : \mathbb{R} \to \mathbb{Z}, \tag{3.27}$$

with the following property

$$\text{quant}_Q(s) \mapsto \left\lfloor \frac{s}{Q} + \frac{1}{2} \right\rfloor, \quad \forall s \in \mathbb{R}.$$

When we consider a vector $x \in \mathbb{R}^n$ that contains all x-coordinates of a model's time step, we apply the quantization quant_Q for every element of this array.

Definition 3.56 (quantum [142]). *We call the step size $Q \in \mathbb{R}$ of a quantization quantum.*

Since we apply the same quantum for all elements of the array $x \in \mathbb{R}^n$, the quantization is called uniform.

For a uniform p-bit quantization with $p \in \mathbb{N}$ we get the following quantization step size

$$Q = 2^{-(p-1)} \left(\max_{i=1,\ldots,n} x_i - \min_{i=1,\ldots,n} x_i \right).$$

We call $p \in \mathbb{N}$ bit per symbol (bps). Dequantization is the process of multiplying quantized values $\text{quant}(x) \in \mathbb{Z}^n$ with its quantum $Q \in \mathbb{R}$. We can see that the reconstruction error $q_x = \text{quant}(x) \cdot Q - x \in \mathbb{R}^n$ is bound by

$$\max_{i=1,\ldots,n} |(q_x)_i| \leq 2^{-p} \left(\max_{i=1,\ldots,n} x_i - \min_{i=1,\ldots,n} x_i \right) = Q/2 \ . \tag{3.28}$$

Therefore, the distance between the original value and the reconstructed one is less or equal to $Q/2$. For further information on quantization errors, we refer to [142]. Figure 3.4 shows a sinusoid in the interval $[0, 2\pi]$ and quantized version with several quantums. For every lossy compression there is a trade-off between loss on information and gain of compression rate. Moreover, for every application field the lossy compression technique has to be adapted and the quantums have to be set individually to fulfill the respective precision requirements. Usually, the utilized technique influences properties of the introduced error. There are sophisticated quantization methods like vector quantization and adaptive quantization [142]. Transformation before a quantization also effects the properties of the error [125, 95]. A specific application field are optimal control problems with nonlinear time-dependent three dimensional PDEs [141, 51, 50, 49]. Here Götschel and Weiser compress state trajectories to limit the demands on storage and bandwidth

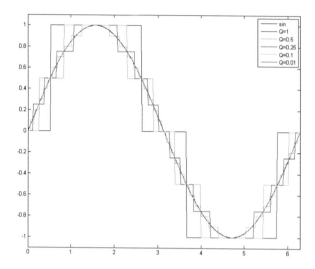

Figure 3.4: Quantized sinusoid with several quanta Q [95]

for keeping state data for solving the adjoint equation. Since solving PDEs numerically results in discretization errors, the authors apply a lossy data compression. To mathematically guarantee the convergence of their investigated optimization, they specify the precision such that they can mathematically guarantee the convergence of the investigated optimization. Based on the problem they apply uniform [141] and adaptive quantization [50].

In the context of computational fluid dynamics [8] the relation between the longest edge between two nodes and the shortest edge can range between several thousand and a few million. For the compression of the coordinates an adaptive precision is meaningful to avoid a degeneration of the mesh [97].

For the compression of crash test simulation results, we have a different situation, since the finite elements usually have a similar size.

3.3.2   Prediction and transformation

In this section we investigate prediction and transformation methods used to improve the properties of data distribution for compression. Efficient prediction methods exploit all available correlations of the investigated data. For simulation of crash tests e.g. correlations exist in space as well as in time [132, 133].

Prediction describes the process of modifying the data in a way that the resulting numbers are both easy to encode and can be reconstructed without any additional loss. A practical and relevant approach is to design the prediction so that the resulting absolute integer numbers are small. In this case, the number of different occurring values and, therefore, the overall used alphabet will be small, too. This can be exploited in the encoding step as shown in the following example.

For example, the prediction eliminates the dependencies between time steps and the topology of a finite element mesh. We use some given values to predict the remaining values. We then determine the residual of the prediction as the difference between the original values and the predicted values. Next the given values and the residual of the prediction will be encoded. For the reconstruction we repeat the prediction and add the residual. For integer data, this procedure is lossless and is reproducible on different machines. The following example demonstrates the importance of prediction methods.

Example. Let S be a stream containing the values $1, 2, ..., 100$ in increasing order. An entropy encoder determines a probability of $P(S_i) = \frac{1}{100}$ for all elements $i = 1, ..., 100$. Therefore, the entropy of the data set would be $H(S) = \sum_{i=1}^{100} \frac{1}{100} \log_2(\frac{1}{100}) = \log_2(\frac{1}{100}) \approx 6.64$ if we assume i.i.d. variables. A dictionary based encoder will not find any reoccurence and has to save all values not encoded. It can exploit that the numbers of 1 to 100 can be described by 7 bits. In both cases the encoding is not very successful, because the encoder cannot recognize the structure of the stream. Markov based entropy encoders , e. g. Prediction by Partial Matching can learn the distribution on-the-fly, but this is also not successful due to the lack of reoccurrences. Since we usually know the structure, we can preprocess the data, e.g. by applying a prediction method to remove dependencies that cannot be exploited by a standard encoder.

When we apply a row prediction P_{row} that predicts each value by its predecessor, except the first one, we will get the following stream:

$$P_{\text{row}}(S) = \{1, 1, ..., 1\}$$

Applying an entropy encoder on the predicted stream will result in a zero entropy, since the probability for the letter "1" is one. Also for the dictionary based method we find the letter "1" in the investigated window. A sophisticated encoder will apply a run length encoding which stores the number 1 once and then its number of occurrences.

As shown in the example a standard encoder cannot exploit all dependencies, that are hidden in the data. Moreover, e.g. for small data sets, it does not make sense to develop an encoder for all types of dependencies. A better strategy is to develop prediction methods and combine them with standard encoders. It is almost always sufficient to reduce the used letters of the alphabet with a prediction and applying a residual coding afterwards. There several prediction techniques applicable on data defined on irregular meshes that we will state briefly in the next sections. Hierarchical methods [141, 132, 133] and wavelet-based methods [49, 107] will not be considered in this thesis.

3.3.2.1 Parallelogram Predictor

A well-known method in the context of data compression of graph-based data is the parallelogram predictor [135, 65] and its generalization for higher dimensions [62], the so-called Lorenzo predictor. These methods suit the compression of data defined on quadrilateral and hexagonal meshes.

The parallelogram predictor was initially applied for the coordinate compression of triangular meshes in the context of 3D meshes [135]. The basic idea is to predict the coordinates of a triangle by those of an already processed triangle.

Let $\{v_1, ..., v_4\} \in V$ a node set, $(v_1, v_2), (v_2, v_3), (v_3, v_1), (v_3, v_4), (v_4, v_2) \in E$ a edge set and

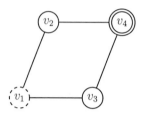

Figure 3.5: *Parallelogram predictor, see Equation (3.29).*

$(x_i, y_i, z_i) \in \mathbb{R}^3$ the coordinates for node v_i with $i = 1, ..., 4$. If we already know a triangle with nodes v_1, v_2, v_3 and edges $(v_1, v_2), (v_2, v_3), (v_3, v_1)$ and we have a triangle v_2, v_3, v_4 which shares the edge (v_2, v_3), we can create a quadrilateral with nodes v_1, v_2, v_3, v_4. Afterwards we can use the coordinate values of v_1, v_2, v_3 to predict those of v_4 in the following way:

$$v_4 = v_3 + v_2 - v_1 \tag{3.29}$$

See Figure 3.5.

The prediction technique is called parallelogram predictor since for a parallelogram in two dimensions as well as an embedded parallelogram in 3D the residual of coordinate prediction for node v_4 is 0 [135]. A proof for the prediction quality as well as a generalization to multidimensional parallelograms, so-called parallelepipeds, is given in the next section.

3.3.2.2   Lorenzo Predictor

The Lorenzo Predictor is a generalization to dimensions $n \in \mathbb{N}\setminus\{0, 1\}$ of the two dimensional parallelogram predictor. It was introduced by Iberria et. al. in [62]. There it was shown that polynomials of degree less than n can be predicted exactly on the corners of an n dimensional unit cube.

Theorem 3.57. *Let the function c be defined as $c : \mathbb{Z}_2^n \rightarrow \{0, ..., n\}$ with $c(u) \mapsto \|u\|_1$. For a given polynomial P in n variables of degree $m < n$, the sum of the signed values $(-1)^{c(u)}P(u)$ over all nodes $u \in \mathbb{Z}_2^n = \{0, 1\}^n$ of the unit cube is zero.*

For a generalization of Theorem 3.57 to parallelepipeds, we need the following notations and definitions. For $f : \mathbb{R}^n \rightarrow \mathbb{R}$, $\mathbf{x} \in \mathbb{R}^n$, and $\mathbf{p_i} \in \mathbb{R}^n\setminus\{\mathbf{0}\}$ for $i = 1, ..., n$, we write

$$d_{\mathbf{p_i}}f(\mathbf{x}) := f(\mathbf{x}) - f(\mathbf{x} - \mathbf{p_i}) \ \forall i = 1, ..., n \tag{3.30}$$

The following lemma describes the commutativity of the operator from Equation (3.30).

Lemma 3.58. *Let $f : \mathbb{R}^n \rightarrow \mathbb{R}$, $\mathbf{x} \in \mathbb{R}^n$ and $\mathbf{p}, \mathbf{q} \in \mathbb{R}^n\setminus\{\mathbf{0}\}$. The difference defined in Equation (3.30) is commutative in respect to operators $d_{\mathbf{p}}$ and $d_{\mathbf{q}}$, i.e. it holds*

$$d_{\mathbf{p}} (d_{\mathbf{q}}f(\mathbf{x})) = d_{\mathbf{q}} (d_{\mathbf{p}}f(\mathbf{x})) .$$

PROOF: Let $f : \mathbb{R}^n \rightarrow \mathbb{R}$, $\mathbf{x} \in \mathbb{R}^n$, and $\mathbf{p}, \mathbf{q} \in \mathbb{R}^n\setminus\{\mathbf{0}\}$ be given.

$$d_{\mathbf{p}} (d_{\mathbf{q}}f(\mathbf{x})) = d_{\mathbf{p}} (f(\mathbf{x}) - f(\mathbf{x} - \mathbf{q}))$$

$$= f(\mathbf{x}) - f(\mathbf{x} - \mathbf{q}) - f(\mathbf{x} - \mathbf{p}) + f(\mathbf{x} - \mathbf{q} - \mathbf{p})$$
$$= (f(\mathbf{x}) - f(\mathbf{x} - \mathbf{p})) - (f(\mathbf{x} - \mathbf{q}) - f(\mathbf{x} - \mathbf{q} - \mathbf{p}))$$
$$= d_{\mathbf{q}}(f(\mathbf{x}) - f(\mathbf{x} - \mathbf{p}))$$
$$= d_{\mathbf{q}}(d_{\mathbf{p}}f(\mathbf{x}))$$

$\square$

We state the definitions of the Lorenzo Predictor and the Lorenzo Difference from [106], which was used for a generalization of Theorem 3.57 to rectangular hexahedrons.

Definition 3.59 (Lorenzo Predictor, Lorenzo Difference [106]). *For* $f : \mathbb{R}^n \to \mathbb{R}$, $\mathbf{x} \in \mathbb{R}^n$, $\mathbf{p_1}, ..., \mathbf{p_n} \in \mathbb{R}^n \backslash \{\mathbf{0}\}$*, we write*

$$(d_{L,\mathbf{p_1},...,\mathbf{p_n}}f)(\mathbf{x}) := \bigotimes_{i=1}^{n} (d_{\mathbf{p_i}}f)(\mathbf{x}) \tag{3.31}$$

$$(L_{\mathbf{p_1},...,\mathbf{p_n}}f)(\mathbf{x}) := f(\mathbf{x}) - (d_{L,\mathbf{p_1},...,\mathbf{p_n}}f)(\mathbf{x}) \ . \tag{3.32}$$

In this context the product sign means the composition of the operators $d_{\mathbf{p_i}}$ *for* $i = 1, ..., n$ *in arbitrarily order, since it is commutative, see Lemma 3.58. We call (3.31) the Lorenzo Difference and (3.32) the Lorenzo Predictor.*

With this notation and the restriction $\mathbf{p_i} = h_i \mathbf{e_i}$, whereas $\mathbf{e_i} \in \mathbb{R}^n$ for $i = 1, ..., n$ are the unit vectors, we can formulate the corollary from [106].

Corollary 3.60 ([106]). *Let* $f : \mathbb{R}^n \to \mathbb{R}$ *be n-times continuously differentiable on the hyperrectangle* $R^{(\mathbf{x})} := \prod_{i=1}^{n}[x_i - h_i, x_i]$*, then the following estimate holds*

$$|(d_{L,h_1\mathbf{e_1},...,h_n\mathbf{e_n}}f)(x_1, ..., x_n)| \leq \prod_{i=1}^{n} h_i \max_{\xi \in R^{(\mathbf{x})}} \left| \frac{\partial^n f(\xi)}{\partial x_1 \cdot ... \cdot \partial x_n} \right| \ . \tag{3.33}$$

The strategy of the proof from [106] can be utilized to obtain a more generalized statement. For functions with values on general parallelepipeds that will be predicted by the Lorenzo Difference.

Definition 3.61 (parallelepiped). *Let* $\mathbf{x} \in \mathbb{R}^n$ *and* $\mathbf{p_1}, ..., \mathbf{p_k} \in \mathbb{R}^n$ *with* $k \leq n$*. We define the k dimensional parallelepiped (PE) embedded in* $\mathbb{R}^n$ *as*

$$P_k^{(\mathbf{x})}(\mathbf{p_1}, ..., \mathbf{p_k}) = \left\{ \mathbf{p} \in \mathbb{R}^n \ \middle| \ \mathbf{p} = \mathbf{x} - \sum_{i=1}^{k} \tilde{a}_i \mathbf{p_i}; \tilde{a}_i \in [0, 1] \right\} \subset \mathbb{R}^n \ . \tag{3.34}$$

Definition 3.62 (derivative, directional derivative [6]). *Let E and F be Banach spaces, $p \in X \backslash \{0\}$ and $f : X \subset E \to F$ be continuously differentiable at $x_0 \in X$. We call $\partial f : X \to \mathcal{L}(E, F)$ derivative of f in x_0, whereas $\mathcal{L}(E, F)$ is the set of bounded linear operators. We call $\frac{\partial f}{\partial p}(x_0) = \lim_{t \to 0} \frac{f(x_0 + tp) - f(x_0)}{t}$ directional derivative.*

With the previous definitions, we can formulate our following theorem.

Theorem 3.63. *Let $f : \mathbb{R}^n \to \mathbb{R}$ be k-times continuously differentiable on the parallelepiped $P_k^{(\mathbf{x})}$ like in (3.34). Then, the following estimate holds*

$$|(d_{L,\mathbf{p_1},...,\mathbf{p_n}} f)(x_1, ..., x_n)| \leq \prod_{i=1}^{n} \|\mathbf{p_i}\|_2 \max_{\xi \in P_k^{(\mathbf{x})}} \left\| \frac{\partial^n f(\xi)}{\partial \mathbf{q_1} \cdot ... \cdot \partial \mathbf{q_n}} \right\|_{\infty} \qquad (3.35)$$

The directions of differentiation are $\mathbf{q_i} = \frac{\mathbf{p_i}}{\|\mathbf{p_i}\|_2}$ for $i = 1, ..., n$.

As a preparation of the theorem's proof, we define a node set of the PE $P_k^{(\mathbf{x})}$.

Definition 3.64 (node set of parallelepiped). *Let the PE be given as in (3.34) and let $\mathbf{x} \in \mathbb{R}^n$. The node set of such a PE is defined as*

$$Y_k^{(\mathbf{x})} = \left\{ \mathbf{y} \in \mathbb{R}^n \,\middle|\, \mathbf{y} = \mathbf{x} - \sum_{i=1}^{k} a_i^{(\mathbf{y})} \mathbf{p_i}; a_i^{(\mathbf{y})} \in \{0, 1\} \right\} \subset \mathbb{R}^n \qquad (3.36)$$

Using this definition, we prove the following Lemma.

Lemma 3.65. *Let $f : \mathbb{R}^n \to \mathbb{R}$ be the function, which describes the values on the nodes of the parallelepiped $P^{(\mathbf{x})}$. These nodes form the set $Y_k^{(\mathbf{x})}$ like in (3.36). For the Lorenzo differences with $c_k(\mathbf{y}) = \sum_{i=1}^{k} a_i^{(\mathbf{y})}$ and $k \leq n$, it holds*

$$(d_{L,\mathbf{p_1},...,\mathbf{p_k}} f)(\mathbf{x}) = f(\mathbf{x}) - \underbrace{\sum_{\mathbf{y} \in Y_k^{(\mathbf{x})} \setminus \{\mathbf{x}\}} (-1)^{c_k(\mathbf{y})+1} f(\mathbf{y})}_{=:\tilde{f}(\mathbf{x})} \ . \qquad (3.37)$$

With the help of this Lemma, we state the proof of Theorem 3.63.

PROOF: The proof uses induction on the dimension k of the PE in the n dimensional vector space $\mathbb{R}^n$.
Let us define $d(\mathbf{x}) := f(\mathbf{x}) - \tilde{f}(\mathbf{x})$.
For $k = 1$: $\mathbf{x} = x_1$

$$d(\mathbf{x}) = f(\mathbf{x}) - \sum_{\mathbf{y} \in Y_1^{(\mathbf{x})} \setminus \{\mathbf{x}\}} (-1)^{c_k(\mathbf{y})+1} f(\mathbf{y})$$

$$= f(x_1) - f(x_1 - \mathbf{p_1})$$

$$= (d_{L,\mathbf{p_1}} f)(\mathbf{x})$$

Assuming the induction hypothesis for dimension $k < n$ holds, i.e. $d(\mathbf{x}) = (d_{L,\mathbf{p_1},...,\mathbf{p_k}} f)(\mathbf{x})$, we show that it also holds for dimension $k + 1$.
Set $\tilde{Y} := \{\mathbf{y} \in Y_{k+1}^{(\mathbf{x})} | a_{k+1}^{(\mathbf{y})} = 0\}$ and $\hat{Y} := \{\mathbf{y} \in Y_{k+1}^{(\mathbf{x})} | a_{k+1}^{(\mathbf{y})} = 1\}$. It is obvious that $\tilde{Y} = Y_k^{(\mathbf{x})}$ and $\hat{Y} = Y_k^{(\mathbf{x} - \mathbf{p_{k+1}})}$ hold.
Moreover, let the function $\hat{c}_k : \mathbb{R}^n \to \mathbb{N}_0$ be $\hat{c}_k(\hat{\mathbf{y}}) = \sum_{i=1}^{k+1} a_i^{(\hat{\mathbf{y}})} - a_{k+1}^{(\hat{\mathbf{y}})}$ for all $\hat{\mathbf{y}} \in \hat{Y}$. The variable $a_{k+1}^{(\hat{\mathbf{y}})}$ is equal to one for all $\hat{\mathbf{y}} \in \hat{Y}$ and $k < n$. It follows that

$$d(\mathbf{x}) = f(\mathbf{x}) - \sum_{\mathbf{y} \in Y_{k+1}^{(\mathbf{x})} \setminus \{\mathbf{x}\}} (-1)^{c_{k+1}(\mathbf{y})+1} f(\mathbf{y})$$

$$= - \sum_{\mathbf{y} \in Y_{k+1}^{(\mathbf{x})}} (-1)^{c_{k+1}(\mathbf{y})+1} f(\mathbf{y})$$

$$= - \sum_{\tilde{\mathbf{y}} \in \tilde{Y}} (-1)^{c_{k+1}(\tilde{\mathbf{y}})+1} f(\tilde{\mathbf{y}}) - \sum_{\hat{\mathbf{y}} \in \hat{Y}} (-1)^{c_{k+1}(\hat{\mathbf{y}})+2} f(\hat{\mathbf{y}})$$

$$= - \sum_{\tilde{\mathbf{y}} \in \tilde{Y}} (-1)^{c_k(\tilde{\mathbf{y}})+1} f(\tilde{\mathbf{y}}) - \sum_{\hat{\mathbf{y}} \in \hat{Y}} (-1)^{\hat{c}_k(\hat{\mathbf{y}})+1} f(\hat{\mathbf{y}})$$

$$= \left(\bigotimes_{i=1}^{k} d_{\mathbf{p}_i} f \right)(\mathbf{x}) - \left(\bigotimes_{i=1}^{k} d_{\mathbf{p}_i} f \right)(\mathbf{x} - \mathbf{p}_{k+1})$$

$$= d_{\mathbf{p}_{k+1}} \left(\left(\bigotimes_{i=1}^{k} d_{\mathbf{p}_i} f \right)(\mathbf{x}) \right)$$

$$= \left(\bigotimes_{i=1}^{k+1} d_{\mathbf{p}_i} f \right)(\mathbf{x})$$

$$= (d_{L,\mathbf{p}_1,...,\mathbf{p}_{k+1}} f)(\mathbf{x}).$$

□

The seventh equation holds, due to Lemma 3.58. For the proof of Theorem 3.63 we additionally need a statement that the directional derivatives can be expressed by a product of derivative and direction vectors.

Theorem 3.66 ([6]). *Let E and F be Banach spaces, $p \in X$ and $f : X \subset E \to F$ be differentiable at $x_0 \in X$. Then there exists a $\frac{\partial f}{\partial p}(x_0)$ for all $p \in X \backslash \{0\}$ with*

$$\frac{\partial f}{\partial p}(x_0) = \partial f(x_0) p \tag{3.38}$$

.

PROOF: We apply the theorem by using the fundamental theorem of calculus.
First we define a curve $\gamma_i(\cdot, \mathbf{y}) : [-\|\mathbf{p}_i\|_2, 0] \to \mathbb{R}^n$ and $\gamma_i(\cdot, \mathbf{y}) : t \mapsto \mathbf{y} + t \frac{\mathbf{p}_i}{\|\mathbf{p}_i\|_2}$ for all $i = 1, ..., n$. The velocity of the curve $\dot{\gamma}_i(\cdot, \mathbf{y})$ is equal to one for all $i = 1, ..., n$. Let $q_i := \frac{\mathbf{p}_i}{\|\mathbf{p}_i\|_2}$ be the normalized direction of p_i for all $i = 1, ..., n$.

$$(d_{L,\mathbf{p}_1,...,\mathbf{p}_n} f)(\mathbf{x})$$

$$= \left(\bigotimes_{i=2}^{n} d_{\mathbf{p}_i} (d_{\mathbf{p}_1} f)(\mathbf{x}) \right)$$

$$= \left(\bigotimes_{i=2}^{n} d_{\mathbf{p}_i} \right)(f(\mathbf{x}) - f(\mathbf{x} - \mathbf{p}_1))$$

$$= \left(\bigotimes_{i=2}^{n} d_{\mathbf{p}_i} \right) \left(\int_{-\|\mathbf{p}_1\|_2}^{0} \frac{d}{dt_1} f(\gamma_1(t_1, \mathbf{x})) \cdot \underbrace{\|\dot{\gamma}_1(t_1, \mathbf{y})\|}_{=1} dt_1 \right)$$

$$= \left(\bigotimes_{i=3}^{n} d_{\mathbf{p_i}} \right) \left(d_{\mathbf{p_2}} \underbrace{ \left(\underbrace{ \int_{-\|\mathbf{p_1}\|_2}^{0} \frac{d}{dt_1} f(\gamma_1(t_1, \mathbf{x})) dt_1 }_{:=\tilde{f}(\mathbf{x})} \right) }_{\tilde{f}(\mathbf{x}) - \tilde{f}(\mathbf{x} - \mathbf{p_2})} \right)$$

$$= \left(\bigotimes_{i=3}^{n} d_{\mathbf{p_i}} \right) \left(\int_{-\|\mathbf{p_1}\|_2}^{0} \frac{d}{dt_1} f(\gamma_1(t_1, \mathbf{x})) dt_1 - \int_{-\|\mathbf{p_1}\|_2}^{0} \frac{d}{dt_1} f(\gamma_1(t_1, \mathbf{x} - \mathbf{p_2})) dt_1 \right)$$

$$= \left(\bigotimes_{i=3}^{n} d_{\mathbf{p_i}} \right) \left(\int_{-\|\mathbf{p_1}\|_2}^{0} \frac{d}{dt_1} \left(f(\gamma_1(t_1, \mathbf{x})) - f(\gamma_1(t_1, \mathbf{x} - \mathbf{p_2})) \right) dt_1 \right)$$

$$= \left(\bigotimes_{i=3}^{n} d_{\mathbf{p_i}} \right) \left(\int_{-\|\mathbf{p_1}\|_2}^{0} \frac{d}{dt_1} \int_{-\|\mathbf{p_2}\|_2}^{0} \left(\frac{d}{dt_2} f(\gamma_2(t_2, \gamma_1(t_1, \mathbf{x}))) \right) dt_2 dt_1 \right)$$

$$= \left(\bigotimes_{i=3}^{n} d_{\mathbf{p_i}} \right) \left(\int_{-\|\mathbf{p_1}\|_2}^{0} \int_{-\|\mathbf{p_2}\|_2}^{0} \left(\frac{d^2}{dt_1 dt_2} f(\gamma_2(t_2, \gamma_1(t_1, \mathbf{x}))) \right) dt_2 dt_1 \right)$$

$$= \left(\int_{-\|\mathbf{p_1}\|_2}^{0} \cdots \int_{-\|\mathbf{p_n}\|_2}^{0} \left(\frac{d^n}{dt_1 \cdots dt_n} f(\gamma_n(t_n, \gamma_{n-1}(t_{n-1}, \gamma_{n-2}(\ldots, \gamma_1(t_1, \mathbf{x}))))) \right) dt_n \cdots dt_1 \right)$$

So far we have used equivalent reformulations. In the next step, we are looking for an upper bound of the Lorenzo differences:

$$| (d_{L,\mathbf{p_1},\ldots,\mathbf{p_n}} f)(\mathbf{x}) |$$

$$\leq \left(\bigotimes_{i=1}^{n} \|\mathbf{p_i}\|_2 \right) \cdot \max_{t_i \in (-\|p_i\|_2, 0), \forall i=1,\ldots,n} \left| \frac{d^n}{dt_1 \cdots dt_n} f(\gamma_n(t_n, \gamma_{n-1}(t_{n-1}, \gamma_{n-2}(\ldots, \gamma_1(t_1, \mathbf{x}))))) \right|$$
$$\underbrace{}_{=\mathbf{x} + \sum_{i=1}^{n} t_i \frac{p_i}{\|p_i\|_2}}$$

$$= \left(\bigotimes_{i=1}^{n} \|\mathbf{p_i}\|_2 \right) \cdot \max_{t_i \in (-\|p_i\|_2, 0), \forall i=1,\ldots,n} \left| \frac{d^{n-1}}{dt_2 \cdots dt_n} f'\left(x + \sum_{i=1}^{n} t_i q_i\right) \cdot \underbrace{\frac{d}{dt_1}\left(x + \sum_{i=1}^{n} t_i q_i\right)}_{q_1} \right|$$

$$= \left(\bigotimes_{i=1}^{n} \|\mathbf{p_i}\|_2 \right) \cdot \max_{t_i \in (-\|p_i\|_2, 0), \forall i=1,\ldots,n} \left| \frac{d^{n-1}}{dt_2 \cdots dt_n} \frac{\partial}{\partial q_1} f\left(x + \sum_{i=1}^{n} t_i q_i\right) \right|$$

$$= \left(\bigotimes_{i=1}^{n} \|\mathbf{p_i}\|_2 \right) \cdot \max_{t_i \in (-\|p_i\|_2, 0), \forall i=1,\ldots,n} \left| \frac{\partial}{\partial q_1} \frac{d^{n-1}}{dt_2 \cdots dt_n} f\left(x + \sum_{i=1}^{n} t_i q_i\right) \right|$$

$$= \left(\bigotimes_{i=1}^{n} \|\mathbf{p_i}\|_2 \right) \cdot \max_{t_i \in (-\|p_i\|_2, 0), \forall i = 1, \dots, n} \left| \frac{\partial^n}{\partial \mathbf{q_1} \cdot \dots \cdot \partial \mathbf{q_n}} f\left(x + \underbrace{\sum_{i=1}^{n} t_i q_i}_{\text{all points in } P^{(x)}} \right) \right|$$

$$= \left(\bigotimes_{i=1}^{n} \|\mathbf{p_i}\|_2 \right) \cdot \max_{\xi \in P^{(x)}} \left| \frac{\partial^n}{\partial q_1 \cdot \dots \cdot \partial q_n} f(\xi) \right|$$

The third equation holds due to Theorem 3.66, the fourth one holds, since f is n-times continuously differentiable. □

In Figure 3.6, we see how the prediction of 2D, 3D, and 4D elements works, if we know all nodes of the element except one. The main challenge is to find a strategy to traverse the grid and process all elements in a way that a high dimensional Lorenzo prediction can be applied as often as possible.

The advantage of the Lorenzo predictor is, that especially for parallelepiped the prediction of node values is very good. The main disadvantage is, that we need a certain regularity and find an order to process all elements efficiently.

3.3.2.3 Tree differences

In this section, we briefly state the tree differences, which can be applied on graph-based data on irregular grids. For further information, we refer to [107, 131].
Let $T = (V, E_t)$ be a tree with a node set $V = \{n_1, \dots, n_n\}$ with $n = \#V$ and an edge set E_t where each edge is oriented in direction towards the leafs of the tree. If we have a variable vector $v \in \mathbb{Z}^n$ with values defined on each node, we will predict all values v_i for all nodes $i \in V \setminus \{n_{\text{root}}\}$ by the value of its predecessor. This reduces the correlation of the data in regard to the tree. The value of the root node cannot be predicted. Figure 3.7 shows an example for the application of tree differences.
The main task when applying tree differences for grids and graphs is to find a traversal. A tree enables the exploiting of dependencies between neighbored nodes while we can ensure that there are no circular dependencies, which would make a decompression impossible. Since we store the connectivities lossless there are two different strategies to determine a spanning tree. On the one hand, we can use a deterministic method that always generates an identical tree if it is applied on the same graph defined by the connectivities. On the other hand, we can store a tree that is optimal for compression. Efficient tree formulations like the condensed tree format [107] are available. The first approach has the advantage of less overhead but with the drawback of a computationally expensive task during decompression. The second approach of storing a tree is usually beneficial if the tree can be used for several data sets like time dependent variables. In such cases the overhead is often neglectable and can even benefit from a special choice of the spanning tree.

3.3.3 Encoding

The final step of a state-of-the-art compression method for simulation results is the encoding of the possibly quantized and predicted data sets. This step is responsible for the reduction in disk space, while the quantization and prediction steps are only applied to

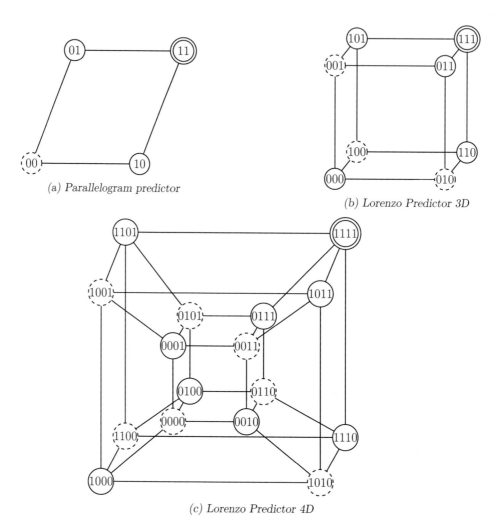

(a) Parallelogram predictor

(b) Lorenzo Predictor 3D

(c) Lorenzo Predictor 4D

Figure 3.6: Lorenzo predictor for a two, three, and four dimensional parallelepiped. The value of the double framed node is predicted by summing up all single framed node values and subtracting all values from dashed framed nodes. The numbers are the node Ids which help to identify if the node value has to be added or subtracted, see $a_i^{(y)} \in 0, 1$ of Equation (3.36)

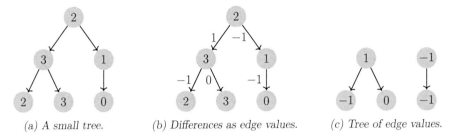

(a) A small tree.   (b) Differences as edge values.   (c) Tree of edge values.

Figure 3.7: Tree differences for a small tree. The tree of edge values can be used for additional applications of tree differences.

modify the properties of the data set to gain better compression rates by the encoder. In Section 3.2, we have categorized various encoding methods. The following descriptions are based on [116].

3.3.3.1 Entropy encoders

Two famous representatives of the entropy encoder family are Huffman and Arithmetic coders [116, 143, 25, 93, 115]. Dynamic Markov Coding, Prediction by partial matching and the Rice encoder [145] also belong to this class. Huffman and Arithmetic coding using the same statistical model, as input but use a different strategy to assign bits to identify the encoded values.

Huffman

The Huffman encoder is one of most used encoders worldwide. It is a famous representative of the class of entropy encoders and is part of the popular lossless compression tool gzip [44] and the compression library zlib [111].
It is based on building a tree of codewords that represents symbols with high probabilities with a short encoded representation and symbols with a low probability with a long representation. The algorithm generates a binary tree, the so called Huffman tree. Every symbol of the alphabet belongs to one leaf and its code can be derived by binary decisions on the way from the root nodes to the leaf.
An example of a Huffman tree for the example in Section 3.1.1 can be found in Figure 3.8. The advantages are that the method itself is simple, efficient and produces the best codes for individual data symbols [25]. The main disadvantage of Huffman coding is that it only produces optimal variable sized codes if the probabilities are powers of two. Moreover, if the statistics change over time or the initial guess of the statistic is not sufficient, there are adaptive and dynamic Huffman encoders, e.g. Faller-Gallager-Knuth and Vitter algorithm. The main drawback of using an adaptive Huffman encoder is that it loses its speed compared to the nonadaptive version. "On the other hand, adaptive minimum-redundancy (Huffman) coding is expensive in both time and memory space, and is handsomely outperformed by adaptive arithmetic coding even ignoring the difference in compression effectiveness" [92].

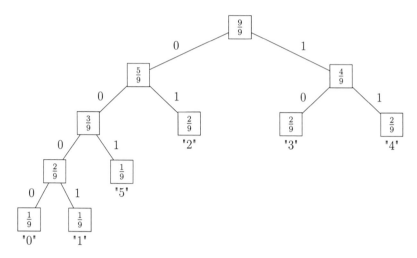

Figure 3.8: *Huffman tree for the example of Section 3.1.1. The code for each letter of the alphabet* $\mathcal{A} = \{0, 1, ..., 5\}$ *can directly be read off the edges starting from the root and going to the respective leaf.*

3.3.3.2 Arithmetic coding

Arithmetic encoding is a statistical compression method that represents a complete input stream with only one number. This is a fundamental difference to Huffman and Shannon-Fano and Rice encoding as they assign certain codewords to elements of the alphabet. The basic idea of arithmetic coding is to represent the entire input stream as a value in the interval $[0, 1)$. The knowledge that the value is nonnegative and smaller than one is used in decompression. Therefore, there is no need to store the leading bit 0 and thus it will be omitted. In the first step the frequencies of occurrence have to be estimated or determined. If the statistics are determined the data has to be analyzed before it will be encoded. Because the data has to be processed twice this kind of encoding is called two-pass compression. Moreover, the arithmetic encoder is a first-in first-out encoder in contrast to the last-in first-out RANS encoder of Duda [34].
After initializing the statistics, the following steps are executed:

1. Definition of the current interval as $[0, 1)$

2. Execution of the following two steps for each symbol s of the input stream:

    (a) Divide the current interval into subintervals whose size are proportional to the symbols' probabilities

    (b) Select the subinterval for s and define it as the current interval

3. When the entire input stream has been processed, the output should be any number that uniquely identifies the current interval $[I_l, I_h)$, i.e. s is identified with the interval $\in [I_l, I_h)$.

The aim is to find the shortest number that uniquely identifies the interval. For each symbol processed, the current interval usually becomes smaller. Therefore, it takes more bits to uniquely identify the interval but it contains the information of all symbols that have been already processed. The major difference to e.g. Huffman coding is, that the final output is a single number and does not consist of codes for individual symbols. The following example describes the arithmetic coding for the input stream of length nine:

$$I = \{2, 0, 2, 3, 4, 4, 5, 3, 1\}. \tag{3.39}$$

The alphabet is $\mathcal{A} = \{0, 1, 2, 3, 4, 5\}$.

Definition 3.67 (relative frequencies, cumulated frequencies). *Let a finite alphabet $\mathcal{A} = \{a_1, ..., a_n\}$, $n \in \mathbb{N}$ and a input stream $I = \{i_1, ..., i_l\} \in \mathcal{A}^l$, $l \in \mathbb{N}$ be given. For a letter $a_i \in \mathcal{A}$, we call $p_{\mathcal{A}}(a_i) = \frac{\#\{a_i \in I\}}{l}$ relative frequency. Moreover, we define the cumulated frequencies as $p_{\mathcal{A}}^{\Sigma}(a_i) = \sum_{j=1}^{i} p_{\mathcal{A}}(a_j)$.*

In the first step we determine the distribution by calculating the relative frequencies, see Table 3.4.

a	$\#$	Relative frequency $p_{\mathcal{A}}(a)$	Cumulated frequencies $p_{\mathcal{A}}^{\Sigma}(a)$
0	1	$\frac{1}{9}$	$\frac{1}{9}$
1	1	$\frac{1}{9}$	$\frac{2}{9}$
2	2	$\frac{2}{9}$	$\frac{4}{9}$
3	2	$\frac{2}{9}$	$\frac{6}{9}$
4	2	$\frac{2}{9}$	$\frac{8}{9}$
5	1	$\frac{1}{9}$	$\frac{9}{9}$
Σ	9	1	

Table 3.4: *Relative and cumulated relative frequencies of the input stream (3.39).*

To determine a new interval we have to take the current one and divide it into sections that represent the distribution in the current step. For this example, the probabilities do not change over time. Therefore, it is a static encoding method. Let I_l and I_u be the lower and upper interval boundary, respectively. Let $p_{\mathcal{A}}$ be the relative frequencies and $p_{\mathcal{A}}^{\Sigma}$ the cumulated relative frequencies from Table 3.4. We determine the new interval boundaries after encoding the n-th symbol of the input stream $s_n \in \mathcal{A} = \{0, ..., 5\}$ with $n \in \mathbb{N} \setminus \{0\}$ by the following rule, assuming that $p_{\mathcal{A}}^{\Sigma}(-1) = 0$:

$$I_l^{n+1} = I_l^n + (I_u^n - I_l^n) \cdot p_{\mathcal{A}}^{\Sigma}(s_n - 1)$$
$$I_u^{n+1} = I_l^n + (I_u^n - I_l^n) \cdot p_{\mathcal{A}}^{\Sigma}(s_n)$$

For the interval after the first step, we get the following lower and upper bounds:

$$I_l^1 = 0 + (1 - 0) \cdot p_{\mathcal{A}}^{\Sigma}(2 - 1) = \frac{2}{3^2}$$
$$I_u^1 = 0 + (1 - 0) \cdot p_{\mathcal{A}}^{\Sigma}(2 - 1) = \frac{4}{3^2}$$

An overview over all bounds can be found in Table 3.5.

n	1	2	3	4	5	6	7	8	9
I_l^n	$\frac{2}{3^2}$	$\frac{18}{3^4}$	$\frac{166}{3^6}$	$\frac{1510}{3^8}$	$\frac{13638}{3^{10}}$	$\frac{122838}{3^{12}}$	$\frac{1105798}{3^{14}}$	$\frac{9952310}{3^{16}}$	$\frac{89570854}{3^{18}}$
I_u^n	$\frac{4}{3^2}$	$\frac{20}{3^4}$	$\frac{170}{3^6}$	$\frac{1518}{3^8}$	$\frac{13654}{3^{10}}$	$\frac{122870}{3^{12}}$	$\frac{1105830}{3^{14}}$	$\frac{9952374}{3^{16}}$	$\frac{89570918}{3^{18}}$

Table 3.5: Lower and upper boundaries of the interval.

The final boundaries are $I_l^9 = \frac{89570854}{3^{18}} \approx 0.2311980304$ and $I_u^9 = \frac{89570918}{3^{18}} \approx 0.2311981956$. A binary number that lies in the interval $[I_l^9; I_u^9)$ independent of its trailing bits is:

$$A = (0.0011101100101111100110)_2$$
$$= 1 \cdot 2^{-3} + 1 \cdot 2^{-4} + 1 \cdot 2^{-5} + 1 \cdot 2^{-7} + 1 \cdot 2^{-8} + 1 \cdot 2^{-11} + 1 \cdot 2^{-13} + 1 \cdot 2^{-14}$$
$$+ 1 \cdot 2^{-15} + 1 \cdot 2^{-16} + 1 \cdot 2^{-17} + 1 \cdot 2^{-18} + 1 \cdot 2^{-21} + 1 \cdot 2^{-22}$$
$$\approx 0.2311980724$$

Even if all following entries of the binary representation of A would be 1, the number will be less than I_u^9. As we can determine all intervals by the given statistic, it is sufficient to store only one value that specifies the actual interval.

Remark 3.68. *The input stream (3.39) can be compressed with an arithmetic encoder using 23 bits, when the decoder has also access to the applied distribution.*

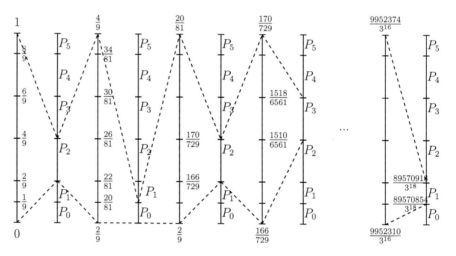

Figure 3.9: Arithmetic coding of stream (3.39): $2, 0, 2, 3, 4, 4, 5, 3, 1$.

The encoded bits are a result of the so called renormalization of the encoding, where the interval is inflated by doubling the size until further redoubling would exceed the maximum interval size of one. The implementation of arithmetic coding is detailed in [116], [12], and [11].

The advantages and disadvantages of the arithmetic coding are summarized in the following remark:

Remark 3.69. *Arithmetic coding has three major advantages:*

- *It is independent of the underlying distributions.*

- *It almost reaches the entropy limit for the compression rate (bits per symbol).*

- *The probabilities used in Step 2a of the workflow description can change in every step without any negative influence on the compression factor.*

The major disadvantage is:

- *The arithmetic encoding and decoding process is slower compared to nonadaptive Huffman coding [114].*

3.3.3.3   Prediction by partial matching

Prediction by partial matching (PPM) is a context-based statistical compression method, that applies an adaptive arithmetic encoder to execute the actual compression. Usually, the context is restricted to a range of 2 to 10 symbols [116]. It is categorized as a logical encoder and was developed by John Cleary and Ian Witten in 1984 [21].

We distinguish between static and adaptive PPM encoders. Static PPM encoders use a finite data set to determine a distribution used for encoding. An occurrence of a value with zero probability can be problematic, since a static arithmetic encoder cannot handle this situation. We modify it by storing such a value without encoding in the compressed output stream. This strategy points in the direction of adaptive PPM encoders.

Adaptive PPM encoders also maintain tables of all possible transitions of the alphabet and uses the tables to assign a probability to the next symbol with regard to the context. The tables are updated continuously, when new symbols will be encoded. Compared to a static version with data probabilities differing from the average, compression rates are higher. The main disadvantage is that this method is more complex and thus slower than a static PPM.

The PPM is a logical encoder. The central idea of PPM is to start with the longest context and if we do not hit a nonzero probability for the next symbol we will decrease the size of the context by one. We will repeat this procedure until we hit a nonzero probability or the value itself has not yet occurred. If it is its first occurrence, the PPM will encode the value using the inverse of the size of the alphabet as the probability.

Since the decoder does not know about the next symbol, we have to store when we reduce the size of the context to encode the next value. Therefore, a so-called escape symbol will be encoded arithmetically and stored in the compressed output stream.

There are several variants of PPM available such as PPMA, PPMB, PPMC, PPMP, and PPMX. The main difference is how they compute the probability for the escape codes. For further information on PPM and its variants, we refer to [116].

3.3.3.4   Rice-Golomb encoding

The Rice encoder is a parametrized prefix code and is a special version of a Golomb encoder [116]. A Golomb encoder is a lossless, adaptive, symmetric, statistical, and logical one-pass encoder, see Section 3.2.

Golomb encoders assume a discrete geometric distribution for the input data stream and can only be applied on nonnegative integer numbers [145]. A Golomb code for a nonnegative integer $b \in \mathbb{N}_0$ for parameter $m \in \mathbb{N}_0$ is determined by evaluating the three quantities $q \in \mathbb{N}_0$, $r \in \mathbb{N}_0$ and $c \in \mathbb{N}_0$ as

$$q = \left\lfloor \frac{b}{m} \right\rfloor, \quad r = b - qm, \text{ and } c = \lceil \log_2 m \rceil. \tag{3.40}$$

For a Rice encoder, parameter m needs to be representable as

$$m = 2^p, p \in \mathbb{N}_0. \tag{3.41}$$

A major advantage of Rice encoding compared to static Huffman encoding is that it can adapt its applied prefix code to the data stream by choosing a parameter $m \in \mathbb{N}_0$. It works with predefined statistical distributions and has a code table for every parameter m. In the compressed data stream the encoded data is stored as well as an identifier for the applied code table.

We use the extended Rice algorithm based on [110, 145], see Figure 3.10. The preprocessing step of extended Rice algorithm contains a bijective mapping $\mathcal{I} : \mathbb{Z} \mapsto \mathbb{N}$:

$$\mathcal{I}(a) = \begin{cases} 2a & \text{for } a \geq 0 \\ 2a - 1 & \text{for } a < 0 \end{cases} \quad \forall a \in \mathbb{Z}. \tag{3.42}$$

The mappings allows us to encode data streams containing negative integers. Let $J \in \{8, 16\}$ be the number elements that are compressed as one block with the same code option. Let $x_1, ..., x_J \in \mathbb{Z}$ be the integer input data and $\delta_1 = \mathcal{I}(x_1), ..., \delta_J = \mathcal{I}(x_J) \in \mathbb{N}_0$ be the preprocessed data. In the case of highly compressible data, the extended Rice encoder applies two special modes. The "Zero block" option is used in the case of $\delta_1 = ... = \delta_J = 0$. If at least one nonzero value δ_i with $i \in \{1, ..., J\}$ exists in the investigated block, the "Second Extension" option is applied. In this case, value tuples δ_{2i-1} and δ_{2i} with $i \in \{1, ..., \frac{J}{2}\}$ are transformed to a single new symbol as follows

$$\gamma_i = \frac{(\delta_{2i-1} + \delta_{2i})(\delta_{2i-1} + \delta_{2i} + 1)}{2} + \delta_{2i}. \tag{3.43}$$

The transformed value γ_i is encoded with a unary code, see [116]:

$$c(\gamma_i) = \begin{cases} 1 & \gamma_i = 0 \\ \underbrace{0...0}_{\text{n-times } 0} 1 & \gamma_i = n \end{cases} \quad \forall i = 1, ..., \frac{J}{2} \tag{3.44}$$

In the case of high entropy data the two special modes do not succeed to compress the block efficiently. In such cases the parameter for an input stream block $\delta_1, ..., \delta_J \in \mathbb{N}_0$ is determined by

$$m = \max_{\mu}\{\mu \leq \frac{1}{J}\sum_{i=1}^{J}\delta_i | \mu = 2^p, p \in \mathbb{N}\},$$

see [106].
A Rice encoder is fast in encoding and decoding and works well if the input stream has a discrete geometric distribution.

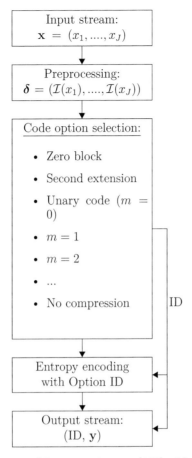

Figure 3.10: Architecture of Rice encoder, see [145] with J being the block size

3.3.3.5 Dynamic Markov coding

Dynamic Markov coding is an adaptive two-stage statistical compression method that applies an adaptive arithmetic encoder in the second stage [23]. The main difference to standard arithmetic coding is, that the distribution is not based on the relative frequencies of values, but determined by a so called finite-state machine [116]. A machine is a model that e.g. counts how often a zero bit is followed by another zero bit or how often changes from zero to one or vice versa will take place. In the first stage, we construct a finite-state machine that estimates a distribution while reading the input stream. In the second stage, we apply an arithmetic encoder that performs the actual compression. The challenge is to determine a finite-state machine that will generate a low entropy of the input stream and achieve the best compression ratio without generating a vast amount of overhead.

In contrast to the PPM, the DMC is a physical encoder, since it usually only takes some bits into account for generating a distribution. Moreover, it is an asymmetric encoder as the data has to be processed twice.

3.3.3.6   Dictionary-based encoders

In this section we briefly introduce dictionary-based encoders. One famous representative is the LZ77 Window method developed by Lempel and Ziv [150]. A variation is integrated in the famous lossless compression tool gzip [44] and the compression library zlib [111]. The family of dictionary based methods are "often used in practice to compress data that cannot be modeled simply, such as English text or computer source code" [25]. In our approach, we also use the implementation of the zlib in two cases. First, if it is not clear which distribution shall be assumed. Second, if we handle small data sets that can be compressed efficiently with the zlib .
For further information we refer to [116].

3.3.3.7   Graphical models

In this section, we state graphical models and the special cases of Bayesian networks and Markov Random fields, which are also used for data compression [43, 27]. The most important application area is machine learning, especially the special case of Bayesian networks, see e.g. [76, 67, 10].
When we consider graph-based data, we can interpret every node as a single random variable, see Section 5.4. Since we want to compress all random variable realizations and assume dependencies between the random variables, the best achievable compression rate is achieved if we use the joint distribution of all random variables in the graph. With the size of the alphabet m and the number of random variables n, respectively, this would result in m^n different events. Therefore, we get a statistic with m^n different elements and just as many codewords. For our use cases with an alphabet size of about 200 and several hundreds of nodes, this would be more than the numbers of atoms in our universe. We need an abstraction and a restriction of the dependencies and determination of independences used for modeling. Graphical models are used for representing conditional independences between a set of random variables and represent a particular factorization of a joint distribution in a graphical manner [45].
For graphical models, we distinguish between graphs with directed edges that leads us to Bayesian Networks and in the case of graphs with undirected edges to Markov Random Fields, see Figure 3.11. In the following two paragraphs, we state a few approaches for

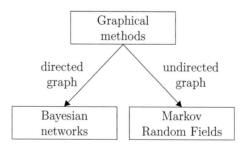

Figure 3.11: Distinction of graphical methods between Bayesian networks and Markov Random Fields.

data compression in the contexts of Bayesian Networks and Markov Random Fields.

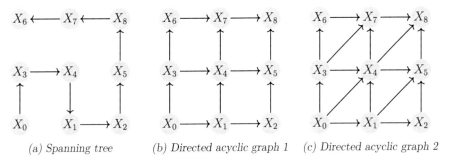

(a) Spanning tree   (b) Directed acyclic graph 1   (c) Directed acyclic graph 2

Figure 3.12: Three directed acyclic graphs (DAG) that occur in different combinations of graph-based prediction methods and iMc encoding, which will be investigated in Section 5.7.2.

Bayesian Networks

We start with the definition of Bayesian networks.

Definition 3.70 (Bayesian Network [67]). *A Bayesian Network is a tuple (X, E) consisting of a set of random variables $X = \{X_0, ..., X_n\}$ and a set of directed edges E between these variables with the following properties. Each variable has a finite set of mutually exclusive states $\mathcal{A} = \{a_0, ..., a_{m-1}\}$. The variables together with the directed edges form a directed acyclic graph (DAG), i.e. a cycle free directed graph. To each random variable X_i with parents $X_{i_1}, ..., X_{i_l}$, there is attached a conditional probability table*

$$\mathbb{P}[X_i | X_{i_1}, ..., X_{i_l}].$$

The main restriction is that the directed edges do not create any cycle [76].

Remark 3.71. *The principle idea of Bayesian networks is to define in a graphical way, which random variables of a set of random variables are conditional independent and if they are not independent, which random variables they are dependent on. The Bayesian network defines what probabilities need to be determined. It does not dictate a certain distribution but only specifies the conditional independences and dependencies of the set of random variables.*

The Bayesian network aims to give a joint distribution for all variables. It states which random variables are independent and helps therefore to speed up the calculation of a joint distribution and managing the statistics. Three examples are displayed in Figure 3.12. In Table 3.6 all conditional probabilities are defined to fully describe the joint distribution of all random variables of $\{X_0, ..., X_8\}$.

The hierarchical structure and relations of graphical probabilistic models are displayed in Figure 3.13.

Remark 3.72. *The Bayesian networks are a generalization of time homogeneous Markov models, to inhomogeneous models up to a model that allows dependencies that are not explicitly represented as an edge.*

| Random | Probabilities for the graph of | | |
variable	Figure 3.12a	Figure 3.12b	Figure 3.12c
X_0	$\mathbb{P}[X_0]$	$\mathbb{P}[X_0]$	$\mathbb{P}[X_0]$
X_1	$\mathbb{P}[X_1\|X_4]$	$\mathbb{P}[X_1\|X_0]$	$\mathbb{P}[X_1\|X_0]$
X_2	$\mathbb{P}[X_2\|X_1]$	$\mathbb{P}[X_2\|X_1]$	$\mathbb{P}[X_2\|X_1]$
X_3	$\mathbb{P}[X_3\|X_0]$	$\mathbb{P}[X_3\|X_0]$	$\mathbb{P}[X_3\|X_0]$
X_4	$\mathbb{P}[X_4\|X_3]$	$\mathbb{P}[X_4\|X_3,X_1]$	$\mathbb{P}[X_4\|X_3,X_1,X_0]$
X_5	$\mathbb{P}[X_5\|X_2]$	$\mathbb{P}[X_5\|X_4,X_2]$	$\mathbb{P}[X_5\|X_4,X_2,X_1]$
X_6	$\mathbb{P}[X_6\|X_7]$	$\mathbb{P}[X_6\|X_3]$	$\mathbb{P}[X_6\|X_3]$
X_7	$\mathbb{P}[X_7\|X_8]$	$\mathbb{P}[X_7\|X_6,X_4]$	$\mathbb{P}[X_7\|X_6,X_4,X_3]$
X_8	$\mathbb{P}[X_8\|X_5]$	$\mathbb{P}[X_8\|X_7,X_5]$	$\mathbb{P}[X_8\|X_7,X_5,X_4]$

Table 3.6: *Dependencies specified by the directed acyclic graphs in Figure 3.12. The mentioned probabilities have to be defined for a complete description of the Bayesian network.*

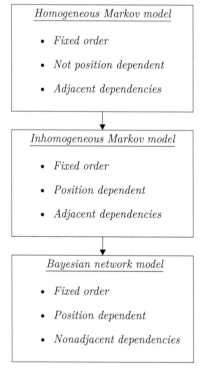

Figure 3.13: *Hierarchical structure of Markovian graphical models, excerpt of [10]. The position can be interpreted as the point in time of a Markov chain compared to a position on a graph in the sense of a graphical model.*

If we remove the restriction of time homogeneity we get an inhomogeneous Markov model. If we further remove the restriction that a node has the maximum of one edge pointing to itself, i. e. we allow the occurrence of DAGs, we get a Bayesian network, see Figure 3.13. A special property of Bayesian networks is that information can also travel from one node to its ancestors. This is the case if we already know the value of a random variable with one of its ancestors still unknown. In this case, the knowledge of this value, also called evidence, influences the distribution of its ancestors in the DAG. There are two ways to reflect such an influence. For singly connected networks, in which the underlying undirected graph has no loops, i.e. a tree, there exists a general algorithm called belief propagation [45]. For multiply connected networks, there exists a more general algorithm known as junction tree algorithm [45]. For more information, we refer to [67, 10].

In [43], Frey investigated the compression of small black-white pictures, namely 6×6 pixel pictures with random horizontal and vertical lines and 8×8 pixel pictures that show hand written digits. In both cases every pixel is a random variable that can embrace a black or white coloring. One interesting result is that he needs a certain number of data sets for learning the distribution before he could outperform gzip compression [44]. For the pictures with lines he needs more than 100 pictures to beat gzip.

In [27], Davies and Moore applied Bayesian networks for the compression of tables with several thousand up to 3 million data sets with 2 to 100 variables per item. Here every variable is represented by a random variable in the Bayesian network. For these data sets the Bayesian networks gain distinctly better compression rates than gzip. In both cases, arithmetic coding is used to actually encode the data.

Bayesian networks are a machine learning approach. For a well-trained distribution, we need quite a big number of data sets. Since no statistics are transmitted the statistics have to be learned on the encoder as well as on the decoder site. This results in two drawbacks.

Remark 3.73. *First, updating the statistics can be computationally expensive in the time-critical decoding phase. Second, if we do not have already compressed all data sets in the same order and learned the statistic like the encoder did, we are not able to decompress the data set. Therefore, there is no random access possible. If we interpret every time step as one training data set for the Bayesian network approach we always have to decompress all previous time steps before we can access the requested one.*

Hidden Markov Model

Hidden Markov models can be seen as a special case of Bayesian networks, see Section 3.3.3.7 [45, 10]. It was developed in the 1960s [9] and is usually used to model time series data.

Definition 3.74 (Hidden Markov Model [45]). *A Hidden Markov Model (HMM) is a tool for representing probability distributions over sequences of observations. On the one hand, it is assumed that the observation at time t was generated by some process whose state X_t is hidden from the observer, see Figure 3.14. On the other hand, we assume that the state of this hidden process satisfies the Markov property, see Equation (3.14).*

The main difference to a normal Markov model is that we cannot observe the random variables that fulfill the Markov property directly.

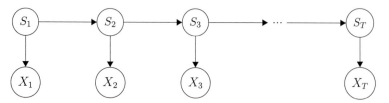

Figure 3.14: Bayesian network of a Hidden Markov Model [45].

Markov Random Fields

Another graphical model are Markov Random Fields (MRF). The main differences of Markov Random Fields to Bayesian Networks are that cyclic dependencies are allowed and the graph is undirected.
We start with the definition of Markov Random Fields.

Definition 3.75 (Markov Random Field [108, 78, 146]). *Let $G = (V, E)$ be an undirected graph with node set V, edge set E, and a set of random variables $X = (X_v)_{v \in V}$. The graph G together with the random variables X forms a Markov Random Field if the measure fulfills the Markov property. I.e. for any two subsets $C_1, C_2 \subset V$ of nodes separated by a third subset U, the random subfields X_{C_1} and X_{C_2} are conditionally independent of each other given the values of X_U*

$$\mathbb{P}[X_{C_1}, X_{C_2}|X_U] = \mathbb{P}[X_{C_1}|X_U]\mathbb{P}[X_{C_2}|X_U].$$

Probabilistic graphical models with undirected edges are generally called Markov Random Fields (MRF) [10].

Remark 3.76. *Identifying a graph and a Markov Random Field works in both directions. If we know the distribution and its conditional independences we can draw the corresponding graph. If we have a graph G we can read off the conditional independences that is the basis for a joint distribution of all nodes of G [101].*

In [108], Reyes investigates the compression of Markov Random Fields. As a practical application he generates black-white (b/w) pictures that follow a homogeneous Ising-model [66]. Since pictures of such a model are "blobby" Reyes also investigates if the Markov Random Field can be used as an underlying distribution for the compression of real world b/w pictures. For such pictures he achieves similar compression rates like the state-of-the-art JBIG encoder [108, 109].
The main difference of his approach compared to [43, 27] is that MRF does not need many data sets to learn a distribution, since the method assumes the Ising model. The compressed data sets are not dependent on previously compressed data sets compared to a machine learning approach. Therefore, random access per image or per time step is possible.
The program sequence is divided in an upward pre-encoding phase, which also has to be done for decoding, and a downward encoding phase. Initially, the root node value is encoded using the distribution determined by the upward phase. In the downward phase the statistics, which are determined in the upward phase, are used next to the value of

the direct ancestor of the node.

A major drawback is the run time complexity, for encoding as well as for decoding.

Remark 3.77. *In the acyclic case, it is dominated by the application of the belief propagation that is $\mathcal{O}((\#\mathcal{A})^2 \#V)$. In the cyclic case, dependent on the used method and if exact interference is necessary, loopy belief propagation or local conditioning are used. Both methods have at least the time complexity of the belief propagation. For the local conditioning the time complexity is usually $\mathcal{O}((\#\mathcal{A})^c \#E)$ with an exponent $c > 2$. Therefore, for cyclic cases the complexity is dependent on the structure of the graph and distinctly higher than in the cyclic case.*

Although there are several successful prediction methods, which can be applied, the alphabet size of the predicted value is usually distinctly higher than 2 for b/w images. Especially the time complexity for decoding is problematic.

For each node, i.e. random variable, there is an own distribution available, which is dependent on a dense matrix of the size $m \times m$ with $m = \#(\mathcal{A})$. In the case of big alphabets and many nodes, the available internal memory can also become a bottle neck.

Additionally, it is not clear which distribution model is meaningfully applicable for simulation results.

Our aim is to use our encoder as a a black box method, that is applicable in both, well and badly predicted data sets. Having discussed the methods from [108], we can exclude this approach and will instead focus on Markov chains and Bayesian networks, see Chapter 5.

Conclusion:

In this chapter we briefly introduce the main concepts of data compression plus components for the compression of simulation results. The focus of this chapter is not to introduce new results but providing the mathematical and methodical background for our new compression techniques Predictive Principal Component Analysis (PPCA), see Chapter 4, and induced Markov chains (iMc), see Chapter 5. Moreover, we gave formal proofs for two statements, namely Theorem 3.42, which says that mean ergodicity is sufficient for a well defined entropy and Theorem 3.63, that gives an error estimation for the Lorenzo prediction for multidimensional parallelepipeds, which are not available in the literature.

We start in Section 3.1 with limits on data compression discussing the concept of entropy for a single random variable and general stochastic processes, see Section 3.1.1 and 3.1.2, respectively. The findings of Sections 3.1.3 and 3.3.3 will provide the basis for our new iMc encoder of Section 5.1. With regard to the classification of our iMc encoder we have stated different types of data compression in Section 3.2. In Section 3.3, we have discussed the main components of data compression for simulation results as well as the compression techniques that are the basis for our proposed methods.

Chapter 4

Dimensionality reduction methods

In this chapter, we tackle the problem how to exploit redundancies between an array of simulation results for data compression with dimensionality reduction methods. For the basic idea of this scientific approach we refer to [90]. In the following, we will provide the mathematical background for the proposed method as well as a theoretical investigation of several efficiency-relevant parameters. As stated in Section 2.3 we have access to a big set of simulation results. We also have access to classification of models to distinguish which simulations can be successfully compressed together. Nonetheless, there are variations in the model geometry where the dimension of node and element variables do not fit together. If only small regions were modified, e. g. one part, a solution to this problem is a mapping of variable values. Another approach is to decompose the car into frequently recurring components, which allows the handling of each part defined by the input deck, see Section 2.2. We opt for the second approach as the information, which parts are available and which are modified is available at SDMS, see Section 2.3. In addition, the decomposition of the whole model allows to express the behavior of the geometry or variables not globally, but locally for each component separately. The principal component analysis (PCA) as dimensionality reduction method is, therefore, not applied on the whole model but on each part of a simulation model. The PCA being a linear method and simulation results in general not being linear we assume the local behavior of single components being closer to linear than the behavior of the entire model. Only for a very small proportion of parts the crash progress can be described by few principal components guaranteeing the required precision. We, therefore, use this method as a prediction method and call it predictive principal component analysis (PPCA). We distinguish between an online and an offline version. The offline method is applied if all simulation results are available at the time of initial compression and is applied if the archiving a set of simulation results is required. The online method is applied if only a small portion of simulation results is available at the time of initial compression. In this case, we presume that all simulation results are available sequentially. This situation is standard for the application of PPCA inside SDMS as only one simulation after another is available and should be compressed immediately, see Remark 2.5.

In contrast to state-of-the-art compression techniques for simulation results, we first apply the prediction and then quantize the residual of the prediction for the PPCA. This approach cannot be applied for general prediction methods as errors of floating point data can accumulate.

The aim of this chapter is to introduce a new compression method that uses techniques of machine learning [4], but does not alter the approximation of previously compressed data sets. The method responds to machine learning by deriving new results if the number of simulation results increases.

We start with an overview of PCA-based compression on three dimensional geometries. Section 4.2 we recollect well-known facts of PCA. In Section 4.3 we introduce our new Offline and Online method of Predictive Principal Component Analysis.

4.1 Previous work

The principal component analysis (PCA) was introduced by Pearson [102] in 1901 and further developed by Hotelling [61] in 1933. In the framework of stochastic processes, PCA was also discovered independently by Karhunen [69] and was generalized by Loève. Therefore, the PCA is also known as Karhunen-Loève transformation.

In the context of time varying geometries with constant mesh connectivities, the application of PCA based methods is fairly new and was first introduced by Lengyel in 1999 [80]. He mentioned that crash test simulations results are a potential source of time varying geometries but does not provide any further details. He discards the application of PCA, since he claims, that PCA cannot capture the relevant information if the change in geometry is described by translations and rotations. Nonetheless, there are several publications applying PCA for data compression, which refer to this paper. In [3], the authors apply a two step preprocessing to overcome the lack of ability to describe a rotation and translation. First they shift the mass center of all vertices into the origin. Second they apply an affine transformation that minimizes the sum over the squared distances between a vertex and its coordinates of the first time step for all vertices. The combination of the shift and the affine transformation is applied to every time step before applying the PCA. For this approach, there was no encoding of the PCA components and the authors used a singular value decomposition calculation on the matrix of size $3n \times T$ with n being the number of nodes and T being the number of time steps. Therefore, they exploit the relation between the coordinate directions. In [70], the PCA method was combined with a quantization of the PCA components and an additional prediction of the coefficients by linear prediction coding, see [116]. Next the PCA base vectors as well as the predicted components are entropy encoded. Additionally, a distortion measurement was introduced. Let A be a $3n \times T$ matrix containing the original animation sequence and $\tilde{A}$ be the compressed and decompressed version of A. Moreover, let $E(A)$ be an average matrix containing the mean vectors for all coordinate directions and all vertices over time. The proposed distortion measure is defined as:

$$e = 100 \frac{\|A - \tilde{A}\|_2}{\|A - E(A)\|_2} \quad . \tag{4.1}$$

Other than state-of-the art compression methods applying uniform quantization PCA methods minimize the reconstruction error in the 2-norm. Additionally PCA is applied if many time steps are available, especially if the number of time steps T exceeds the number of nodes n.

In [117], the combination of PCA with a clustering application is introduced as clustered PCA (CPCA). Additionally, the matrix is of dimension $n \times 3T$ and contains in every third line the coordinates in x, y, and z direction. The nodes are clustered depending on the behavior of their trajectories in time. If they are similar they are part of one cluster. To reduce the computational costs especially in the case of big models with only a few time steps, the authors implement the PCA by an eigenvalue decomposition of the matrix $A^T A$ in the case of $n > 3T$ and on AA^T otherwise.

[7] applies a clustering method as a preprocessing step combining it with a local coordinate systems for each segmentation. It is called Relative Local Principal Component Analysis (RLPCA). The authors apply a quantization on coefficients and base vectors. Compared to CPCA compression is rather slow as a transformation for all coordinates has to be applied.

In [138], the PCA is applied on the trajectories. The spatial coherences remaining in the coefficients are eliminated by using a more sophisticated prediction method than in previous approaches, namely the parallelogram predictor.

[139] focuses on the compression of the PCA base vectors. Vasa and Skala apply nonuniform quantization and non-least-squares optimal linear prediction on the PCA base vectors. In [137], a parallelogram predictor, see Section 3.3.2.1, is applied on the base vectors with a new traversal strategy.

In [112], the authors apply high order singular value decomposition. They decomposed the tensor with the dimension number of vertices times frames times coordinate directions. As reconstruction errors result inappropriate if a reduction of the coordinate directions is executed the authors limit the dimensionality reduction to the number of frames and the number of time steps. Applying the Tucker decomposition the core tensor does not have to be diagonal. Therefore, although no dimensionality reduction regarding the number of coordinate directions takes place, the method can be seen as a generalized singular value decomposition (SVD).

The application of PCA-based compression on simulation results is already known. It was initially suggested by [80]. The first application on simulation results was performed in the context of molecular dynamics by [77]. They combine a PCA as dimensionality reduction method with a discrete cosine transformation and a quantization afterwards. Due to the quantization in the frequency domain [116], the properties of the quantization errors are in a way that their effect on the postprocessing is minimal.

In our approach, we use the result of the PCA method as a prediction method instead of a classic dimensionality reduction approach. Our approach differs from the methods described so far in several respects. The requirements on the precision will be much higher than for 3D graphics by a factor of 10-100. We use the the maximum norm as an error measure, see Chapter 2 as the distortion measure (4.1) is not suitable for limiting reconstruction errors for unique vertices. We will handle each coordinate direction separately. Furthermore, we do a clustering regarding the segmentation of the car in parts. A question is if the segmentation is fine enough or even too fine for certain parts.

The most important challenge will be to determine the inner dimension of the dimensionality reduction method for compression. While all papers applied trial and error strategies we provide a strategy to select them in an almost optimal way.

4.2 Principal component analysis

In this section, we present a brief overview of the most important facts for the principal component analysis, which can be found e.g. in [79]. The core question of a PCA on time dependent data, where each time step is an observation, is if we can uncouple time-dependencies and space dependencies. Considering several models simultaneously we also answer the question if uncoupling of one model can serve as uncoupling for another model. We start with the data model used and its consequential properties.

4.2.1 Data model

The data model of PCA is based on the assumption that variables $(y_1, y_2, ..., y_n)^T \in \mathbb{R}^n$ are a result of a linear transformation W of latent variables $(x_1, ..., x_k)^T \in \mathbb{R}^k$:

$$y = Wx \tag{4.2}$$

Here the number k is the intrinsic dimension of the described model. For the PCA model we assume that all latent variables have a Gaussian distribution and all columns of W are orthonormal. The transformation W is an axis change with an additional mapping in a higher dimensional space. Due to the orthogonal and normalized column vectors of W it holds

$$W^T W = I_k \tag{4.3}$$

Additionally, the PCA model requires centralized latent variables x_j, $j = 1, ..., k$ as well as observed variables y_i, $i = 1, ..., n$.

$$\sum_{i=1}^{n} y_i = 0 \Rightarrow E_y[y] = 0 \tag{4.4}$$

$$\sum_{i=1}^{k} x_i = 0 \Rightarrow E_x[x] = 0 \tag{4.5}$$

This requirement is easily accomplished by subtracting the mean value for each node as an estimator for the average.

The challenge of dimensionality reduction is to determine the latent variables $x_1, ..., x_k$ by observing a few outcomes of the data model $y^{(1)}, ..., y^{(t)}$ with the matrix W being unknown. The PCA solves this task by an application of a singular value decomposition on matrix $A = \left(y^{(1)}, ..., y^{(t)}\right) \in \mathbb{R}^{n \times t}$.

4.2.2 Singular value decomposition

We briefly state the singular value decomposition (SVD) as a central step of the PCA, see [79], and the truncated version we use in our PPCA approach.

Definition 4.1 (Singular value decomposition (SVD), singular value [79, 48]). *The SVD of a matrix $A \in \mathbb{R}^{n \times t}$ is written as*

$$A = U\Sigma V^T , \tag{4.6}$$

where $U \in \mathbb{R}^{n \times n}$ and $V \in \mathbb{R}^{t \times t}$ are orthogonal matrices and $\Sigma \in \mathbb{R}^{n \times t}$ is a pseudo-diagonal matrix. The positive diagonal elements $\sigma_1, ..., \sigma_k$ of Σ are called singular values of A and are sorted in descending order, i.e.

$$\sigma_1 \geq ... \geq \sigma_k > 0$$

Theorem 4.2 ([129]). *The singular value decomposition exists for all complex and real valued matrices.*

When it is applied to complex-valued matrices, U and V are unitary matrices.
Applying a SVD we get a decomposition for the matrix of observations $A = \left(y^{(1)}, ..., y^{(t)}\right) = U\Sigma V^T \in \mathbb{R}^{n \times t}$. Since we want to achieve a dimensionality reduction to $p < \min\{n, t\}$ and k is usually too big (for details see Section 4.3.4.1) we approximate matrix A by a truncated SVD

$$\tilde{A} = U \begin{pmatrix} \text{diag}(\sigma_1, ..., \sigma_p) & 0 \\ 0 & 0 \end{pmatrix} V^T \in \mathbb{R}^{n \times t}$$

$$= U|_{n \times p}\text{diag}(\sigma_1, ..., \sigma_p) \left(V|_{t \times p}\right)^T \in \mathbb{R}^{n \times t}$$

The approximation consists of p principal components that are the columns of $U|_{n \times p}$, the singular values $\sigma_1, ..., \sigma_p$, and the coefficients that are the columns of $V|_{t \times s}$. As a direct consequence, the rank of matrix $\tilde{A}$ is p if $p \leq k$.

4.2.3 Properties

The principal component analysis (PCA) is a linear method fulfilling the requirements of the data model specified in Subsection 4.2.1 as well as the following two criteria, see [79, 80]:

- minimal reconstruction error in 2-norm

- maximal preserved variance and decorrelation.

Both properties can be proven by applying the properties of the SVD, see [79, 80].
In the following let $A \in \mathbb{R}^{n \times t}$ be a rank k matrix with a SVD like in Equation (4.6). The two given properties are valid for the complete matrix $W \in \mathbb{R}^{n \times k}$ with $k \leq t$ from Equation (4.2) and for $W|_{n \times p} \in \mathbb{R}^{n \times p} \, \forall p = 1, ..., k$. We can use the same method with identical properties for a truncated SVD where we only investigate an approximation.

Remark 4.3. *The property of PCA to find the optimal subspace for all dimensions $p = 1, ..., k$ in the sense of minimal reconstruction error and maximal variance preservation is important for our application. Since we apply a PCA on data that shows highly nonlinear behavior, see Remark 2.4, we will likely overestimate the intrinsic dimension. Therefore, getting the best linear subspace for $p \in \{1, ..., k - 1\}$ with $\sigma_{p+1} > 0$ is crucial.*

As a consequence of the reduction to a subspace of dimension $p < k$ the reconstruction error $\|A - \tilde{A}\|_2^2$ is greater than zero.
We overestimate the intrinsic dimension and cut after a certain number of principal components and determine p and accept a nonzero reconstruction error. In [122], Shawe-Taylor

et al. used a SVD on a data matrix $A \in \mathbb{R}^{n \times t}$ to prove several properties of the Gram matrix $G = A^T A$. A byproduct is the formulation of each eigenvalue of $G \in \mathbb{R}^{t \times t}$ as the projected columns of A to its corresponding eigenvector. This follows directly by applying the Courant-Fischer Min-max theorem [36] and by G being positive semi-definite and having a full set of orthogonal eigenvectors which creates an orthonormal basis of $\mathbb{R}^t$ due to the symmetry of matrix G. Let $a_j \in \mathbb{R}^t$ for $j \in 1, ..., n$ be the rows of A, $V \subset \mathbb{R}^t$, and $P_V : \mathbb{R}^t \longrightarrow \mathbb{R}^t$ the projection onto the subspace V. If we consider the i-th eigenvector v_i, it holds [122]:

$$\lambda_i(G) = \sum_{j=1}^{n} \|P_{v_i}(a_j)\|_2^2 \tag{4.7}$$

and with $V_p = \text{span}\{v_1, ..., v_p\}$, $a = P_{V_p}(a) + P_{V_p}^\perp(a)$ for all $a \in \mathbb{R}^t$, and the assumption that $\lambda_1 \geq \lambda_2 \geq ... \geq \lambda_t$:

$$\sum_{i=1}^{p} \lambda_i(G) = \sum_{j=1}^{n} \|a_j\|_2^2 - \sum_{j=1}^{n} \|P_{V_p}^\perp(a_j)\|_2^2, \quad \forall p \in \{1, ..., t\} \tag{4.8}$$

As in [122] it follows that:

$$\sum_{i=1}^{p} \lambda_i(G) = \sum_{j=1}^{n} \|a_j\|_2^2 - \min_{\dim(V)=p} \sum_{j=1}^{n} \|P_V^\perp(a_j)\|_2^2 \tag{4.9}$$

$$\sum_{i=p+1}^{t} \lambda_i(G) = \min_{\dim(V)=k} \sum_{j=1}^{n} \|P_V^\perp(a_j)\|_2^2 . \tag{4.10}$$

With $\lambda_i(G)$ for $i = 1, ..., t$ being the squared singular values of matrix A and the selected eigenspace being determined by an optimization problem the restriction to p columns of W will also result in a minimal reconstruction error in the case of p principal components. Applying a SVD results in a decoupling of time and spatial relation by a linear method.

4.2.4 Calculation of truncated singular value decompositions

In this subsection we discuss, how to determine a truncated singular value decomposition [18]. As we use the PCA only as an approximation for the prediction of the original matrix A and we furthermore, do a lossy data compression, we do not focus on the precision of the singular value decomposition of a centralized matrix $\bar{A}$. We try to approximate a highly nonlinear behavior of a crash test by a linear method, see Remark 4.3. For our application we opt for an approximation that slightly overestimates the intrinsic dimension but contains most information of the data. With the focus being on fast calculation of the SVD we calculate the truncated singular value decomposition via an eigenvalue decomposition (EVD) of the Gram matrix $G = A^T A$ or the covariance matrix $C = AA^T$. Our decision which one to use is similar to [117] but our matrix dimensions are $n \times n$ and $T \times T$ in contrast to $n \times n$ and $3T \times 3T$. Due to machine precision, we introduce an error by calculating the matrix-matrix product G on a computer. The question is, how big this error will be and how big the influence on the singular value decomposition will be. Davis and Kahan showed in [19] that the influence of perturbation on the eigenspaces and eigenvectors is dependent on the eigengaps, i.e. the absolute difference between the eigenvalues. As we want to compute a truncated SVD with intrinsic dimension s

it is sufficient to do a truncated EVD of G with intrinsic dimension s. The difference of the eigenspace determined on G and on $\tilde{G}$ that additionally contains the numerical errors of the matrix multiplication can be bound by the difference of $\sigma_s^2 - \sigma_{s+1}^2$. Let $\{\sigma_1^2, ..., \sigma_k^2, 0, ..., 0\}$ be the spectrum of matrix G. The smaller the difference of $\sigma_s - \sigma_{s+1}$ for $\sigma_s < 0.5$,

$$\sigma_s^2 - \sigma_{s+1}^2 = (\sigma_s - \sigma_{s+1})(\underbrace{\sigma_s}_{<0.5} + \underbrace{\sigma_{s+1}}_{\leq \sigma_s}),$$

becomes even smaller. Thus, using an truncated EVD for the calculation of a truncated SVD, only the biggest singular values has to be considered. This also holds for our approach. After determining the matrix with the leading s eigenvectors of G written as a matrix $V|_{t \times s}$, we use the product

$$A(V|_{t \times s}, 0)diag(\sigma_1^{-1}, ..., \sigma_s^{-1}, 0, ..., 0)$$

to approximate the eigenvectors of the covariance matrix C. Round off errors due to finite precision of a computer can occur. Since V consists of orthonormal vectors and we only consider the reciprocals of the biggest singular values this calculation is stable, at least if the EVD was stable. A consequence of investigating the Gram matrix G instead of A itself is, that the numerical error of determining singular vectors assigned to singular values bigger than 1 is more precise if the calculation of G is exact. For singular values smaller than 1 the investigation of the Gram matrix can lead to instabilities.

4.3 Predictive principal component analysis

In this section, we introduce the PPCA and distinguish between an offline method similar to the approach of [80], and introduce the new online method that is predestinated for application in SDMS. We analyze the PPCA method by e.g. giving a point wise bound on the residual of PPCA prediction, see Theorem 4.6. Moreover, we discuss how to set the parameters such as internal precision and number of principal components that are crucial for a successful compression. Finally, we discuss, how to decompose a simulation model into parts.

We want to use the PPCA Online as well as the PPCA Offline methods as a black box tool except for a user given precision Q. The online method is initialized by applying the offline method on all available simulation results.

An important property of the PPCA is, that we split the elements of PCA in data sets: data sets that can be reused and data sets that are simulation result specific. The data sets that can be reused are stored in a database db file and the simulation specific data sets in a ssf file. This approach is important for the online method.

The main steps of the PPCA Offline method are:

1. decomposition of all available models into parts

2. save the initial coordinates and the connectivities of all parts in file db

3. extract the time dependent data for each parts separately

4. apply a PCA, see Section 4.3.2 and 4.3.3

5. compress the principal components, coefficients, and mean values lossy with an internal precision

6. approximate the original data set by lossy compressed and decompressed principal components, coefficients, and mean values and calculate the residual

7. compress the residual with a user given precision

8. store the compressed principal components and mean values in the *db* file and the compressed coefficients and the residual in a simulation specific file *ssf*.

For each simulation result obtained we check for all parts if they were already processed at the PPCA Offline method. If not yet processed we apply a PPCA Offline method. Otherwise we proceed with the PPCA Online method:

1. subtract the mean values stored in *db* from the new data matrix

2. approximate the data matrix with the principal components stored in *db*

3. apply amendment steps to improve the approximation by eliminating errors due to missing orthogonality

4. compress the new coefficients and store them in file *ssf*

5. determine residual matrix; if entries of residual matrix are efficiently compressible we are done; if not proceed with the following steps

6. apply a PCA on the residual, see Section 4.3.2 and 4.3.3.

7. compress the new principal components and additional coefficients lossy with an internal precision.

8. approximate the residual by lossy compressed and decompressed principal components and coefficients and calculate the new residual.

9. compress the new residual with a user given precision.

10. store the new compressed principal components in a new database file *adb* and the compressed additional coefficients and the new residual in a simulation specific file *ssf*.

Points 1 to 3 of the PPCA Offline method are part of the gathering information step, which will be explained next. In Section 4.3.2, we introduce the PPCA Offline method and give a point wise upper limit on the absolute residual of the prediction. The PPCA Online method is introduced in Section 4.3.3. Here we focus on the amendment steps as a fast way of eliminating errors due to quantization errors that effects the orthogonality of the principal components. In Section 4.3.4, we investigate four parameters of the definition of the PPCA method and influence the overall compression rate as well as the run time. We conclude with a discussion on how to decompose a simulation result into parts.

4.3.1   Gathering information

In this section, we briefly describe what data streams shall be treated by the PPCA and how we collect them in matrices.

We aim to compress geometry and variable data of s simulation runs. Therefore, we collect the data for each simulation result in submatrices $A_1, ..., A_s$ with $A_i \in \mathbb{R}^{n \times t_i}$ for $i = 1, ..., s$. Each column of A_i consists of a data stream at a certain time point $t_i^{(j)}$ for $j = 1, ..., t_i$. In geometry data like coordinates, the data streams consists of one dimension of the displacements for all node data with respect to the initial geometry and n is the number of nodes. For variable data like stresses or strains the data streams consist of the variable data. We cannot subtract initial values in general as we do not want to save them separately and they are more likely to vary in a set of simulations. Moreover, a variable can also be defined on an element of the simulation. In such a case n would be the number of elements.

In general, the data streams do not contain all available data but certain parts of data, e.g. only data of one component of a car. These parts are either given by the simulation result file or generated by a cluster method. In Section 4.3.5, we further investigate which types of parts are possible.

For the sake of simplicity, we restrict ourselves to node-based variables in the following. Let a part $\mathcal{P}$ with n nodes be given. Moreover, $\mathcal{P}$ occurs in simulations $j_1, ..., j_{\hat{s}}$. For a simultaneous processing of the data we gather all submatrices A_i for $i = j_1, ..., j_{\hat{s}}$ into one matrix

$$A = (A_{j_1} A_{j_2} ... A_{j_{\hat{s}}}) \in \mathbb{R}^{n \times t}$$

with $t = \sum_{i=j_1,...,j_{\hat{s}}} t_i$. Due to the PCA requirement of centralized variables, see Section 4.2.3, we have to subtract the mean values of the nodes over all observations $m_j^A = \sum_{i=1}^{t} A_{ji}$ for $j = 1, ..., n$:

$$\bar{A} = A - \left(m_1^A, ..., m_n^A\right)^T \cdot (1, ..., 1) \in \mathbb{R}^{n \times t}.$$

Matrix $\bar{A}$ is the input for our PPCA method that is explained in the next two sections.

4.3.2   Offline method

In this section, we introduce the PPCA Offline method and depict how we calculate the low dimensional approximation via a truncated SVD [18] with optimized computational costs. We illustrate the method for one part since all parts can be handled independently. This additionally enables parallel processing of parts. Furthermore we give an upper bound on the residual of the prediction by lossy compressed PCA elements.

We distinguish two cases for a fast determination of a truncated SVD. In the first case, the number of nodes is not bigger than the number of all time steps: $n \le t$. The centralized matrix $\bar{A}$ has a SVD

$$\bar{A} = V \Sigma U^T,$$

where $V \in \mathbb{R}^{n \times n}$ and $U \in \mathbb{R}^{t \times t}$ are orthogonal matrices and $\Sigma \in \mathbb{R}^{n \times t}$ is a pseudo-diagonal $n \times t$ matrix whose first n columns contain $diag\{\sigma_1, ..., \sigma_n\}$, with ordered diagonal entries $\sigma_1 \ge ... \ge \sigma_n$, and whose last $t - n$ columns are all zero, see Section 4.2.2. For $n \le t$, we calculate an unscaled version of the covariance matrix by

$$C = \bar{A}\bar{A}^T = V\Sigma U^T U\Sigma^T V^T = V\Sigma\Sigma^T V^T \in \mathbb{R}^{n \times n}.$$

Afterwards, we apply the FEAST Eigensolver [103] available in the Intel MKL [64] to get the eigenvectors of C regarding the p biggest nonzero eigenvalues. This can be interpreted as a truncated eigenvalue decomposition (EVD). In the FEAST method, we can additionally specify the region, where we want to look for eigenvalues and how many eigenvalues should be determined. Moreover, the method terminates automatically, if the eigenvalues are smaller than a given threshold. Since we are interested in the singular vectors of $\bar{A}$ corresponding to the biggest p singular values, we accept the instabilities occurring in the eigenspaces corresponding to small eigenvalues by determining the SVD via the covariance matrix for the gain of speed, see Section 4.2.4. The singular values of $\bar{A}$ are the square roots of the nonzero eigenvalues of C. Let $V|_{n \times p} \in \mathbb{R}^{n \times p}$ be the submatrix of V consisting of the first p columns and $\Sigma|_{p \times p} = diag\{\sigma_1, ..., \sigma_p\} \in \mathbb{R}^{p \times p}$. Similar to [126], we get a truncated SVD from the truncated EVD of

$$\tilde{C} = V|_{n \times p}\Sigma|_{p \times p}(\Sigma|_{p \times p})^T(V|_{n \times p})^T,$$

with the inverse $(\Sigma|_{p \times p})^{-1} = diag\{\sigma_1^{-1}, ..., \sigma_p^{-1}\} \in \mathbb{R}^{p \times p}$ and the diagonal matrix $\Sigma|^{t-p \times t-p} = diag(\sigma_{p+1}, ..., \sigma_t)$ by

$$
\begin{aligned}
(U|_{t \times p})^T &= I_{p \times t}U^T \\
&= (\Sigma|_{p \times p})^{-1}\Sigma|_{p \times p}I_{p \times t}U^T \\
&= (\Sigma|_{p \times p})^{-1}(\Sigma|_{p \times p}\ 0)U^T \\
&= (\Sigma|_{p \times p})^{-1}(I_{p \times t}\ 0)\begin{pmatrix} I_{p \times t} \\ 0 \end{pmatrix}(\Sigma|_{p \times p}\ 0)U^T \\
&= (\Sigma|_{p \times p})^{-1}(I_{p \times t}\ 0)\left(\begin{array}{c|c} I_{p \times t} & 0 \\ \hline 0 & 0 \end{array}\right)\left(\begin{array}{c|c} \Sigma|_{p \times p} & 0 \\ \hline 0 & 0 \end{array}\right)U^T \\
&= (\Sigma|_{p \times p})^{-1}(I_{p \times t}\ 0)\left(\begin{array}{c|c} I_{p \times t} & 0 \\ \hline 0 & 0 \end{array}\right)\left(\begin{array}{c|c} \Sigma|_{p \times p} & 0 \\ \hline 0 & \Sigma|^{t-p \times t-p} \end{array}\right)U^T \\
&= (\Sigma|_{p \times p})^{-1}(I_{p \times t}\ 0)\left(\begin{array}{c|c} I_{p \times t} & 0 \\ \hline 0 & 0 \end{array}\right)V^T V\Sigma U^T \\
&= (\Sigma|_{p \times p})^{-1}\underbrace{(I_{p \times t}\ 0)\left(\begin{array}{c|c} I_{p \times t} & 0 \\ \hline 0 & 0 \end{array}\right)V^T}_{(V_{n \times p})^T}\bar{A} \\
&= (\Sigma|_{p \times p})^{-1}(V|_{n \times p})^T\bar{A}.
\end{aligned}
$$

For $n > t$, we apply the EVD on the Gram matrix $\hat{G} = \bar{A}^T\bar{A}$. This approach is part of the linear multidimensional scaling (MDS) methods [80]. We use the same techniques as in the first case to obtain the SVD. For the determination of the first columns of V we do a truncated EVD of

$$G = \bar{A}^T\bar{A} = U\Sigma^T V^T V\Sigma U^T = U\Sigma^T\Sigma U \in \mathbb{R}^{t \times t}.$$

Analogous we get the missing singular vectors by the following calculation

$$
\begin{aligned}
(V|_{n\times p}) &= VI_{n\times p} \\
&= V\Sigma|_{n\times p}(\Sigma|_{p\times p})^{-1} \\
&= V\Sigma U^T U|_{t\times p}\Sigma|_{p\times p})^{-1} \\
&= \bar{A}U|_{t\times p}(\Sigma|_{p\times p})^{-1}.
\end{aligned}
$$

For both cases, we call the first p columns of V, which belong to the biggest singular values, principal components and the first p rows of U^T coefficients. We can usually determine all components of the SVD via a fast symmetric EVD of a low dimensional matrix, a matrix multiplication and scaling. Since we apply a lossy compression and additionally store the residual of the prediction, we can further reduce the number of used principal components. Crucial for a successful application of the PPCA is how to choose the number of used principal components $P \leq p$. Since simulations of crash tests are a nonlinear problem, cf. Section 2.2, and we apply a linear dimensionality reduction method, cf. Section 4.2.3, there, in general, is no indicator such as a spectral gap that indicates the intrinsic dimension [80] and hence the number of needed principal components P. Hence, we apply an optimization process with an objective function being the overall compression rate, see Section 4.3.4.1.

Notation 4.4. *The exponent $\cdot^{\lfloor q \rfloor}$ signalizes that all scalar elements are quantized with quantum q and all numbers are integers.*

Using the PCA elements as a prediction, we have to determine an approximation for the data stream matrix where we apply the quantized mean values

$$
\left(m^A\right)^{\lfloor q \rfloor} = \left(\left(m_1^A\right)^{\lfloor q \rfloor}, ..., \left(m_n^A\right)^{\lfloor q \rfloor}\right)^T \in \mathbb{Z}^n \tag{4.11}
$$

and the quantized elements of the truncated singular value decomposition

$$
\hat{A}^{\lfloor q \rfloor} = \left(V|_{n\times P}(\Sigma|_{P\times P})^{\frac{1}{2}}\right)^{\lfloor q \rfloor} \left(\left(U|_{t\times P}(\Sigma|_{P\times P})^{\frac{1}{2}}\right)^{\lfloor q \rfloor}\right)^T, \tag{4.12}
$$

with $\hat{A}^{\lfloor q \rfloor} \in \mathbb{Z}^{n\times t}$ and $(\Sigma|_{P\times P})^{\frac{1}{2}} = \mathrm{diag}(\sqrt{\sigma_1}, ..., \sqrt{\sigma_P}) \in \mathbb{R}^{P\times P}$.

Remark 4.5. *We scale the principal components as well as the coefficients by the square root of the respective singular value. Applying a uniform quantization on the scaled values, the overall quantization is adaptive. We only consider column vectors of matrix $\left(V|_{n\times P}(\Sigma|_{P\times P})^{\frac{1}{2}}\right)^{\lfloor q \rfloor}$ with length unequal 0. For every column that is equal to zero, we reduce P by one and eliminate the principal component from $\left(V|_{n\times P}(\Sigma|_{P\times P})^{\frac{1}{2}}\right)^{\lfloor q \rfloor}$.*

In the following, we focus on the size of the residual of PPCA Offline prediction and give an upper bound for its absolute biggest element. The residual matrix of the PPCA prediction is

$$
E = A - q \cdot \left(m^A\right)^{\lfloor q \rfloor} \cdot (1, ..., 1) - q^2 \cdot \hat{A}^{\lfloor q \rfloor} \in \mathbb{R}^{n\times t}, \tag{4.13}
$$

where "$\cdot$" defines a scalar multiplication, i.e the uniform dequantization for each entry separately, see Section 3.3.1. We uniformly quantize the elements of E with the user defined precision Q to maintain a maximal error of $\frac{Q}{2}$ for the reconstruction. The elements of residual matrix E consists of errors due to dimensionality reduction via truncated SVD as well as the quantization of mean values, principal components, and coefficients.

$$
\begin{aligned}
E =& A - q \cdot \left(m^A\right)^{\lfloor q\rfloor} \cdot (1, ..., 1) - q^2 \cdot \hat{A}^{\lfloor q\rfloor} \\
=& \underbrace{\left(m_1^A, ..., m_n^A\right)^T \cdot (1, ..., 1) + V\Sigma U^T}_{=A} \underbrace{-V|_{n\times P}\Sigma|_{P\times P}(U|_{t\times P})^T + V|_{n\times P}\Sigma|_{P\times P}(U|_{t\times P})^T}_{=0} \\
& - q \cdot \left(\left(m_1^A\right)^{\lfloor q\rfloor}, ..., \left(m_n^A\right)^{\lfloor q\rfloor}\right)^T \cdot (1, ..., 1) \\
& + q^2 \cdot \left(V|_{n\times P}(\Sigma|_{P\times P})^{\frac{1}{2}}\right)^{\lfloor q\rfloor}\left(\left(U|_{t\times P}(\Sigma|_{P\times P})^{\frac{1}{2}}\right)^{\lfloor q\rfloor}\right)^T \\
=& \underbrace{V\Sigma U^T - V|_{n\times P}\Sigma|_{P\times P}(U|_{t\times P})^T}_{=:R_{\mathrm{DR}}} \\
& + \underbrace{\left(m_1^A, ..., m_n^A\right)^T \cdot (1, ..., 1) - q \cdot \left(\left(m_1^A\right)^{\lfloor q\rfloor}, ..., \left(m_n^A\right)^{\lfloor q\rfloor}\right)^T \cdot (1, ..., 1)}_{=:R_{\mathrm{QM}}} \\
& + \underbrace{V|_{n\times P}\Sigma|_{P\times P}(U|_{t\times P})^T - q^2 \cdot \left(V|_{n\times P}(\Sigma|_{P\times P})^{\frac{1}{2}}\right)^{\lfloor q\rfloor}\left(\left(U|_{t\times P}(\Sigma|_{P\times P})^{\frac{1}{2}}\right)^{\lfloor q\rfloor}\right)^T}_{=:R_{\mathrm{QPC}}} \\
=& R_{\mathrm{DR}} + R_{\mathrm{QM}} + R_{\mathrm{QPC}}
\end{aligned}
\tag{4.14}
$$

The residual is in general unequal to zero due to the following three reasons:

- dimensionality reduction: R_{DR},

- quantization of mean values: R_{QM}, and

- quantization of principal components and coefficients: R_{QPC}.

Since we compress the elements of matrix E and apply a uniform quantization and an entropy encoding with our iMc encoder, see Section 5.1, we are interested in the maximal entries of E.

Theorem 4.6. *Let $A \in \mathbb{R}^{n\times t}$ be a data stream matrix with singular values $\sigma_1, ..., \sigma_s$ with $s \leq \min\{n, t\}$ and a quantum q be given. Then it holds for the residual of the PPCA Offline method when using P principal components that*

$$
|E_{ij}| \leq \sigma_{P+1} + q\left(\frac{1}{2} + \sqrt{\sigma_1 P}\right) + P\left(\frac{q}{2}\right)^2
\tag{4.15}
$$

for all $i = 1, ..., n$ and $j = 1, ..., t$.

PROOF: From (4.14), we know that the residual consists of three elements. The spectral norm of R_{DR} coincides with the biggest singular value due to definition. We can directly read out the biggest singular value with the following observation:

$$
R_{\mathrm{DR}} = V\Sigma U^T - V|_{n\times P}\Sigma|_{P\times P}(U|_{t\times P})^T
$$

$$= V \begin{pmatrix} \Sigma|_{s \times s} & 0 \\ 0 & 0 \end{pmatrix} U^T - V \begin{pmatrix} \Sigma|_{P \times P} & 0 \\ 0 & 0 \end{pmatrix} U^T$$

$$= V \begin{pmatrix} 0 & 0 & 0 \\ 0 & \mathrm{diag}(\sigma_{P+1}, ..., \sigma_s) & 0 \\ 0 & 0 & 0 \end{pmatrix} U^T \qquad (4.16)$$

Except for sorting (4.16) is a singular value decomposition of R_{DR} with σ_{P+1} as the biggest singular value. For all $i = 1, ..., n$ and $j = 1, ..., t$, it holds with basic properties of the Euclidean norm, see [48, 120], that

$$|(R_{\mathrm{DR}})_{ij}| = |e_i^T R_{\mathrm{DR}} e_j| \le \|e_i\|_2 \|R_{\mathrm{DR}} e_j\|_2 \le \|e_i\|_2 \|R_{\mathrm{DR}}\|_2 \|e_j\|_2 = \|R_{\mathrm{DR}}\|_2 \le \sigma_{P+1}. \quad (4.17)$$

We resume with the quantization error of the mean values. From Section 3.3.1, we know that the maximal error for a uniform quantization with quantum q can be bound by $\frac{q}{2}$. Therefore, for all $i = 1, ..., n$ and $j = 1, ..., t$, it holds that

$$|(R_{\mathrm{QM}})_{ij}| \le \frac{q}{2}. \qquad (4.18)$$

We finish with the investigation of R_{QPC}. Here the error can accumulate due to summations. Let

$$F^V := q \cdot \left(V|_{n \times P} (\Sigma|_{P \times P})^{\frac{1}{2}} \right)^{\lfloor q \rfloor} - V|_{n \times P} (\Sigma|_{P \times P})^{\frac{1}{2}}$$

be the matrix of quantization errors for the scaled principal components and

$$F^U := q \cdot \left(U|_{t \times P} (\Sigma|_{P \times P})^{\frac{1}{2}} \right)^{\lfloor q \rfloor} - U|_{t \times P} (\Sigma|_{P \times P})^{\frac{1}{2}}$$

be the matrix of quantization errors for the scaled coefficients. Let $(V_{ik})_{k=1,...,P} \in \mathbb{R}^P$ the i-th row vector of V, containing its first P elements. Since matrices V and U of the SVD are orthogonal, also the row vectors have Euclidean length 1. Together with the equivalence of norms $\| \cdot \|_1$ and $\| \cdot \|_2$ [120, 48] we have

$$\sum_{k=1}^{P} |V_{ik}| = \|(V_{ik})_{k=1,...,P}\|_1 \le \sqrt{P} \|(V_{ik})_{k=1,...,P}\|_2 \le \sqrt{P} \|(V_{ik})_{k=1,...,N}\|_2 \le \sqrt{P}. \quad (4.19)$$

Analogously, we get

$$\sum_{k=1}^{P} |U_{jk}| \le \sqrt{P}. \qquad (4.20)$$

For all $i = 1, ..., n$ and $j = 1, ..., t$, we have

$$|(R_{\mathrm{QPC}})_{ij}|$$

$$= \left| \left(V|_{n \times P} \Sigma|_{P \times P} (U|_{t \times P})^T - q^2 \cdot \left(V|_{n \times P} (\Sigma|_{P \times P})^{\frac{1}{2}} \right)^{\lfloor q \rfloor} \left(\left(U|_{t \times P} (\Sigma|_{P \times P})^{\frac{1}{2}} \right)^{\lfloor q \rfloor} \right)^T \right)_{ij} \right|$$

$$= \left| \sum_{k=1}^{P} V_{ik} \sqrt{\sigma_k} \sqrt{\sigma_k} U_{jk} - \sum_{k=1}^{P} (V_{ik} \sqrt{\sigma_k} + F_{ik}^V)(\sqrt{\sigma_k} U_{jk} + F_{jk}^U) \right|$$

$$= \left| -\sum_{k=1}^{P} V_{ik} \sqrt{\sigma_k} F_{jk}^U - \sum_{k=1}^{P} F_{ik}^V \sqrt{\sigma_k} U_{jk} - \sum_{k=1}^{P} F_{ik}^V F_{jk}^U \right|$$

$$\leq \sum_{k=1}^{P} |V_{ik} \underbrace{\sqrt{\sigma_k}}_{\leq \sqrt{\sigma_1}} \underbrace{F_{jk}^U}_{\leq \frac{q}{2}}| + \sum_{k=1}^{P} |\underbrace{F_{ik}^V}_{\leq \frac{q}{2}} \underbrace{\sqrt{\sigma_k}}_{\leq \sqrt{\sigma_1}} U_{jk}| + \sum_{k=1}^{P} |\underbrace{F_{ik}^V}_{\leq \frac{q}{2}} \underbrace{F_{jk}^U}_{\leq \frac{q}{2}}|$$

$$\leq \frac{q}{2} \sqrt{\sigma_1} \underbrace{\sum_{k=1}^{P} |V_{ik}|}_{\substack{(4.19) \\ \leq \sqrt{P}}} + \frac{q}{2} \sqrt{\sigma_1} \underbrace{\sum_{k=1}^{P} |U_{jk}|}_{\substack{(4.20) \\ \leq \sqrt{P}}} + P \left(\frac{q}{2}\right)^2$$

$$= q\sqrt{\sigma_1 P} + P\left(\frac{q}{2}\right)^2. \tag{4.21}$$

Utilizing (4.17), (4.18), and (4.21) in (4.14) proofs the statement.   □

The bound in Theorem 4.6 is not sharp since we use estimations for norm equivalences and gauge the first P singular values by the first one. After determining the residual of the PPCA Offline prediction we quantize the residual matrix with the user defined precision Q and encode the result E^Q e. g. with our iMc encoder, see Section 5.1. We store the encoded matrix E^Q in the simulation specific file *ssf*. Furthermore, we store the chosen principal components and the mean values in the database *db* and add the coefficients in the *ssf*. An overview of the complete PPCA Offline compression workflow can be found in Figure 1.1.

The reconstruction A_{recon} of data stream matrix A as part of the decompression is determined by

$$A_{\text{recon}} = q \cdot \left(m^A\right)^{\lfloor q \rfloor} \cdot (1, ..., 1) + q^2 \cdot \hat{A}^{\lfloor q \rfloor} + Q \cdot E^Q. \tag{4.22}$$

The decompression is fast since it consists only of summations, scalar multiplications and one low dimensional matrix multiplication.

4.3.3   Online method

In this section, we introduce the iteratively applicable PPCA Online method. It perfectly fits into SDMS as a specialized compression that exploits the big number of simulation results that are available sequential, see Remark 2.5. The initialization of PPCA Online method consists of one application of PPCA Offline method on all previously available simulation results, see Section 4.3.2. Therefore, for the remainder of this section, we assume that we handle a data stream B with respect to a part $\mathcal{P}$ that was already processed by the PPCA Offline method on a data stream A. A new data stream B can be one or more extra simulations run as well as newly generated points in time from a known simulation run.

For all simulations that are executed subsequently we want to use the principal components already calculated as a basis to represent new time steps and observations, respectively.

A major difference to state-of-the-art methods applied in machine learning is that we do not alter the initially calculated basis, since we do not want to update previously compressed files. An update of previously compressed files would be computationally and logistically expensive whereas the expected benefit would be small since the basis was chosen optimal for the data sets available so far.

Let $B \in \mathbb{R}^{n \times r}$ be a new data stream of one or more extra simulation runs with r time steps. We say that two parts are identical if the connectivities and all coordinates of the first time step are identical, see Section 4.3.1. For the compression of the coordinates we consider B as the displacements, analogous to Section 4.3.2.

The majority of parts behaves similar in all simulations. Therefore, the overhead of saving a new mean value vector is generally higher than the overhead of a PCA generated with disturbed mean values. This especially is the case if only one simulation is added per application of the PPCA Online method. Moreover, the goal is to describe several simulations in one coordinate system that is specified by previous simulations. The principal components can be interpreted as axis directions of a new coordinate system. If we would change the origin, in general, these axis directions are not meaningful. Therefore we subtract the dequantized mean values $q \cdot \left(m_1^A\right)^{\lfloor q \rfloor}, ..., q \cdot \left(m_n^A\right)^{\lfloor q \rfloor} \in \mathbb{R}$ of the nodes in A, which are saved in the db file, from matrix B:

$$\bar{B} = B - \left(\left(m_1^A\right)^{\lfloor q \rfloor}, ..., \left(m_n^A\right)^{\lfloor q \rfloor}\right)^T \cdot \underbrace{(q, ..., q)}_{\in \mathbb{R}^r} \in \mathbb{R}^{n \times r}.$$

In the next step, we use the subspace that is spanned by the principal components that are stored in the db file and find the projection of the columns of B into that subspace. For this purpose, we dequantize and normalize the lossy compressed principal components. Let

$$(\tilde{v}_1, ..., \tilde{v}_P) = q \cdot \left(V|_{n \times P}(\Sigma|_{P \times P})^{\frac{1}{2}}\right)^{\lfloor q \rfloor} \in \mathbb{R}^{n \times P} \tag{4.23}$$

with column vectors $\tilde{v}_1, ..., \tilde{v}_P$. Since we neglect all scaled and quantized principal components of zero length by the PPCA Offline step, see Remark 4.5, we have $\|\tilde{v}_i\|_2 \neq 0$ for all $i = 1, ..., P$. We define

$$\hat{V} := (\hat{v}_1, ..., \hat{v}_P) = q \cdot \left(V|_{n \times P}(\Sigma|_{P \times P})^{\frac{1}{2}}\right)^{\lfloor q \rfloor} \operatorname{diag}\left(\frac{1}{\|\tilde{v}_1\|_2}, ..., \frac{1}{\|\tilde{v}_P\|_2}\right) \in \mathbb{R}^{n \times P}. \tag{4.24}$$

We represent the columns of $\bar{B}$ for all $i = 1, ..., r$ by the dequantized and renormalized principal components $\{\hat{v}_1, ..., \hat{v}_P\}$ of $\bar{A}$, i.e. the column vectors of $\hat{V}$ that are stored in the database and $\hat{v}_i = \frac{\tilde{v}_i}{\|\tilde{v}_i\|_2}$ for all $i \in \{1, ..., P\}$. We obtain the corresponding coefficients as a matrix

$$C^{(0)} := \hat{V}^T \bar{B} \in \mathbb{R}^{P \times r}.$$

Now we are able to approximate $\bar{B}$ by

$$B^{(0)} := \hat{V} C^{(0)}.$$

The idea of calculating the coefficients in this way is related to the SVD of matrix $\bar{B} = \tilde{W}\tilde{\Sigma}\tilde{U}^T$. We expect that the subspace that is spanned by the first P columns of $\tilde{W}$ is similar to the subspace of the first P columns of V. We will show in Chapter 6, that these expectations are valid for most of the investigated data sets. Exceptions occur due to the complex process of a crash simulation.

In the PPCA Offline method, principal components are quantized in the compression phase with a quantum q. Two problems can occur:

1. Column vectors of $V|_{n \times P}^{\lfloor q \rfloor}$ are not orthogonal.

2. Euclidean norm of column vectors from $qV|_{n \times P}^{\lfloor q \rfloor}$ is not 1.

The second case is solved by the normalization of the dequantized principal components in Equation (4.24), since we had to eliminate the scaling with the square root of the singular value anyways. The following example illustrates the need for a normalization in a very coarse quantization case.

Remark 4.7. Let be $n = 4$, $P = 1$, and $V|_{n \times P} = (0.5, 0.5, 0.5, 0.5)^T$. If we apply an absolute quantization on $V|_{n \times P}$ with quantum $q = 1$, this results in $V|_{n \times P}^{\lfloor q \rfloor} = (1, 1, 1, 1)^T$. The matrix-matrix product $(V|_{n \times P}^{\lfloor q \rfloor})^T V|_{n \times P}^{\lfloor q \rfloor} = 4$, obviously has the eigenvalue 4 once, which would lead to a divergence of the amendment steps that is introduced next, see also Section 4.3.4.2.

For all parts, we calculate the residual of representing $\bar{B}$ by the P principal components of A.

$$E^{(0)} := \bar{B} - B^{(0)} \in \mathbb{R}^{n \times r}.$$

The basic idea of the PPCA Online method is to apply an SVD on the residual matrix to get further principal components that are not yet available. Therefore, such new principal components describe the difference between the already processed simulations and the new ones. A direct application of the PCA on the residual is not meaningful due to the missing orthogonality of the quantized principal components. This can lead to a situation, where we find singular vectors in $E^{(0)}$ that are linear dependent on the principal components of $\bar{A}$. Bottom line, if we want to approximate the time steps of the new simulation results without redundancy, we have to do either a reorthogonalization or a projection. The computational costs of applying an orthogonalization method like the Gram-Schmidt process, see [48], are very high and would also have to be applied in the decompression phase. Therefore, we use an iterative projection of the residual onto the subspace of $\mathbb{R}^n$ that is spanned by the dequantized and normalized principal components $\{\hat{v}_1, ..., \hat{v}_P\}$ of $\bar{A}$. We call one projection an *amendment step* (AS) that is defined by

$$C^{(k)} = \hat{V}^T E^{(k-1)} + C^{(k-1)} \text{ with } B^{(i)} := \hat{V} C^{(i)} \text{ and } E^{(i)} := \bar{B} - B^{(i)} \; \forall i = 0, ..., k-1. \quad (4.25)$$

If the principal components are orthogonal, $\hat{V} E^{(0)}$ would be a zero matrix. In our application, it is generally nonzero. Therefore, we add the newly calculated representation of the residual as an update for the coefficients to the ones calculated in the first step and determine the residual with the updated coefficients for each amendment step k. Its functionality is visualized in Figure 4.1.

Remark 4.8 ([140]). *The amendment steps are a simultaneous application of the Richardson-Iteration, see [71], on the normal equation, see [48],*

$$\hat{V}^T \hat{V} c_*^{(i)} = \hat{V}^T b^{(i)} \quad (4.26)$$

to solve the following least square problem for every observation $b^{(i)}$ with $\bar{B} = (b^{(1)}, ..., b^{(r)})$

$$\min_{c^{(i)} \in \mathbb{R}^P} \|b^{(i)} - \hat{V} c^{(i)}\|_2. \quad (4.27)$$

The minimum is assumed for $c_*^{(i)}$ that is the i-th column of matrix C^* that contains the optimal coefficients for the representation of $b^{(i)}$, for $i = 1, ..., r$, with $\{\hat{v}_1, ..., \hat{v}_P\}$.

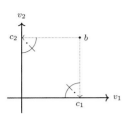

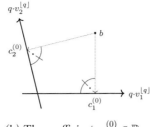

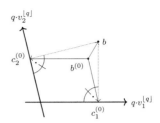

(a) Basis vectors v_1 and v_2 and corresponding coefficients c_1 and c_2.

(b) The coefficients $c_1^{(0)} \in \mathbb{R}$ and $c_2^{(0)} \in \mathbb{R}$ are the $q \cdot v_1^{\lfloor q \rfloor}$ and $q \cdot v_2^{\lfloor q \rfloor}$-intercept.

(c) Due to the missing orthogonality, the reconstruction $b^{(0)} = q(c_1^{(0)} \cdot v_1^{\lfloor q \rfloor} + c_2^{(0)} \cdot v_2^{\lfloor q \rfloor})$ yield an error.

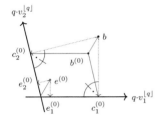

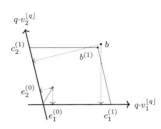

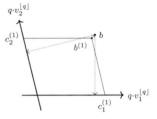

(d) Like vector for b, we perform a orthogonal projection of error $e^{(0)} = b - b^{(0)}$ onto $q \cdot v_1^{\lfloor q \rfloor}$ and $q \cdot v_2^{\lfloor q \rfloor}$.

(e) The coefficients are updated $c_j^{(1)} = c_j^{(0)} + e_j^{(0)}$ for $j = 1, 2$ and the reconstruction $b^{(1)} = q(c_1^{(1)} \cdot v_1^{\lfloor q \rfloor} + c_2^{(1)} \cdot v_2^{\lfloor q \rfloor})$ is performed.

(f) The reconstruction error $e^{(1)} = b - b^{(1)}$ is determined.

Figure 4.1: Effect of one amendment step for the representation of vector $b = c_1 \cdot v_1 + c_2 \cdot v_2 \in \mathbb{R}^2$ with coefficients $c_j^{(i)} \in \mathbb{R}$, $i = 0, 1$, $j = 1, 2$ and vectors $v_1^{\lfloor q \rfloor}, v_2^{\lfloor q \rfloor} \in \mathbb{R}^2$ that are no longer orthogonal due to quantization. The black dotted lines display orthogonal projections onto the vectors v_1, v_2 and $v_1^{\lfloor q \rfloor}, v_2^{\lfloor q \rfloor}$, respectively. The blue lines represent the reconstruction. The red lines show the reconstruction error.

If the precision is not too coarse, the eigenvalues of the Richardson iteration matrix $N = (I - \hat{V}^T \hat{V})$ are less than one. A detailed analysis on the spectrum of $\hat{V}^T \hat{V}$ and a sufficient bound on the quantum q to guarantee convergence for the amendment steps can be found in Section 4.3.4.2.

We apply the amendment steps to improve the mapping on the low dimensional space and keep the error due to dimensionality reduction. The following theorem shows how the amendment steps can be expressed by an iteration matrix $K = (I - H)$.

Theorem 4.9. Let $i \in \mathbb{N}$ and $H := \hat{V} \hat{V}^T$. The error

$$E^{(i)} = \bar{B} - B^{(i)} \text{ with } B^{(i)} = \hat{V} C^{(i)}$$

in the i-th amendment step can be expressed like:

$$E^{(i)} = (I - H)^i E^{(0)} \qquad (4.28)$$

PROOF: Let $i \in \mathbb{N}$, $H = \hat{V}\hat{V}^T$, and $E^{(i)} := \bar{B} - B^{(i)}$ be given.

$$E^{(i+1)} = \bar{B} - \hat{V}C^{(i+1)}$$
$$= \bar{B} - \hat{V}(\hat{V}^T E^{(i)} + C^{(i)})$$
$$= \bar{B} - HE^{(i)} - \underbrace{\hat{V}C^{(i)}}_{=B^{(i)}}$$
$$= \underbrace{\bar{B} - B^{(i)}}_{=E^{(i)}} - HE^{(i)}$$
$$= (I - H)E^{(i)} \tag{4.29}$$

The theorem directly follows by an induction over the number of iterations $i \in \mathbb{N}$.   □

For a closer look on the spectrum of iteration matrix $K = (I - H)$ in regard to the quantum q, see Lemma A.14 and Section 4.3.4.2.

Notation 4.10. *If a capital index of a matrix is equal to P it stands for selecting the first P columns of the matrix. If it is $n - P$ it stands for selecting the last $n - P$ columns.*

Let

$$Q_P^q = V_P - \hat{V}$$

be the quantization error of the P principal components.

$$I - H = VV^T - \hat{V}\hat{V}^T$$
$$= (V_P, V_{n-P})(V_P, V_{n-P})^T - (\hat{V}, 0_{n-P})(\hat{V}, V_{n-P})^T$$
$$= \underbrace{(V_P, V_{n-P})(Q_P^q, 0_{n-P})^T - (V_P, V_{n-P})(Q_P^q, 0_{n-P})^T}_{=0} + (V_P, V_{n-P})(V_P, V_{n-P})^T$$
$$\quad - (\hat{V}, 0_{n-P})(\hat{V}, V_{n-P})^T$$
$$= (V_P, V_{n-P})(Q_P^q, 0_{n-P})^T + (V_P, V_{n-P})(\hat{V}, V_{n-P})^T - (\hat{V}, 0_{n-P})(\hat{V}, V_{n-P})^T$$
$$= (V_P, V_{n-P})(Q_P^q, 0_{n-P})^T + \underbrace{(V_P - \hat{V}, V_{n-P})}_{=Q_P^q}(\hat{V}, V_{n-P})^T$$
$$= (V_P, 0_{n-P})(Q_P^q, 0_{n-P})^T + (Q_P^q, V_{n-P})(\hat{V}, V_{n-P})^T$$
$$= \underbrace{V_P(Q_P^q)^T + Q_P^q \hat{V}^T}_{\text{Elements due to projection on non-orthonormal vectors}} + \underbrace{V \left(\begin{array}{c|c} 0 & 0 \\ \hline 0 & I_{n-P} \end{array} \right) V^T}_{\text{Dimensionality reduction}} \tag{4.30}$$

Applying Equations (4.29) and (4.30) the remaining error after an amendment step results from:

- projection on non-orthonormal vectors

- dimensionality reduction and underestimation of the intrinsic dimension.

If the quantization error equals to zero the following equation holds for all $i \in \mathbb{N}\backslash\{0\}$, see Equation (4.30):

$$K^i := (I - H)^i = V \left(\begin{array}{c|c} 0 & 0 \\ \hline 0 & I_{n-P} \end{array} \right) V^T. \tag{4.31}$$

With no quantization on $V|_{n\times P}$ and infinite machine precision the eigenvalues of the iteration matrix K are $n - P$ times 1 and P times 0. Having no infinite machine precision and applying an absolute quantization on the principal components results in the spectrum being different. For detailed analysis and quantization precisions that guarantee convergence we refer to Section 4.3.4.2.

For the remainder of this section, we expect that the error in direction of the first P eigenvectors vanishes with an increasing number of amendment steps. Before we quantize the coefficients we scale them with the length of the dequantized principal components, that are stored in the database db, see (4.24):

$$\hat{C}^{(i)} := \text{diag}\left(\frac{1}{\|\tilde{v}_1\|_2}, ..., \frac{1}{\|\tilde{v}_P\|_2}\right) C^{(i)} \ \forall i \in \mathbb{N}. \tag{4.32}$$

This is necessary to perform a direct reconstruction without a scaling of the principal components. Afterwards we uniformly quantize $\hat{C}$ with the internal precision q and get $(\hat{C})^{\lfloor q \rfloor}$. We get an approximation for $\bar{B}$ that is

$$\hat{B}^{(i)} := q \cdot \left(\left(V|_{n\times P}(\Sigma|_{P\times P})^{\frac{1}{2}}\right)^{\lfloor q \rfloor} (\hat{C}^{(i)})^{\lfloor q \rfloor})\right) \forall i \in \mathbb{N}.$$

The residual of this approximation is equal to the residual of the reconstruction for all $i \in \mathbb{N}$:

$$F = B - \left(\hat{B}^{(i)} - \left(\left(m_1^A\right)^q, ..., \left(m_n^A\right)^q\right)^T \cdot (q, ..., q)\right)$$

$$= \left(B - \left(\left(m_1^A\right)^q, ..., \left(m_n^A\right)^q\right)^T \cdot (q, ..., q)\right) - \hat{B}^{(i)}$$

$$= \bar{B} - \hat{B}^{(i)}. \tag{4.33}$$

After the application of a amendment steps, there are two possible cases. In the first case, the matrix of displacements $\bar{B}$ is similar to $\hat{B}$ and, therefore, it can easily be represented by the already chosen principal components by encoding the quantized coefficients $(\hat{C})^{\lfloor q \rfloor}$ and storing them in a simulation specific file ssf. In the second case, the representation of $\bar{B}$ by the principal components $\{v_1^{\lfloor q \rfloor}, ..., v_P^{\lfloor q \rfloor}\}$ fails and results in a big residual F. We apply a truncated SVD to the residual F:

$$F \approx V|_{n\times p} \Xi|_{p\times p} (\mathcal{U}|_{r\times p})^T, \text{ with } p < \min\{n, r\}, p \in \mathbb{N}$$

We neglect the centralization. Having to store mean values for every application of the PPCA Online phase would cause a disproportionate overhead. Same as in the PPCA Offline method of Section 4.3.2, we decide how many principal components $\hat{P}$ are used by an optimization process, where the objective function is equal to the overall compression rate, see Section 4.3.4.1. All further steps in regard to the SVD components of residual matrix F are analogous to the Offline method in Section 4.3.2. We add the new $\hat{P}$ lossy compressed principal components of $E^{(a)}$, i.e. the scaled and quantized column vectors of $\mathcal{V}$, to an additional database file adb. Since we diminished the influence of the vectors $\{\hat{v}_1, ..., \hat{v}_P\}$ on the error F by several amendment steps we expect that the $\hat{P}$ newly chosen principal components $\hat{v}_{P+1}, ..., \hat{v}_{P+\hat{P}}$ are at least almost orthogonal to $span\{\hat{v}_1, ..., \hat{v}_P\}$, i. e.

$$0 \leq \langle \hat{v}_i | \hat{v}_j \rangle \ll 1 \ \ \forall i \in \{1, ..., P\}, j \in \{P+1, ..., P+\hat{P}\}.$$

The amendment steps are again applied to fix the missing orthogonality of the quantized principal components. To reduce run time we check based on the Frobenius norm [48] of F the user given precision Q and a threshold δ, if an application of a truncated SVD on F is meaningful.

Databases db and adb and the simulation results specific files ssf are needed for decompression. If we apply the PPCA Online method iteratively, we get a chain of dependencies of several adb files and one db file. It is sufficient to store the information of the previous adb or db file but a certain mechanism has to provide all dependent files for decompression. A SDMS can offer such a mechanism, see Section 2.3.

4.3.4 Crucial parameters

In this section, we investigate four parameters of PPCA Offline and Online method that are crucial for a good compression rate. We face a certain conflict of objectives namely compression rates for certain elements, the overall compression rate, as well as the run time.

- Number of principal components

- Internal precision of mean values, principal components, and coefficients

- Lower bound for the number of nodes of a part to apply the PPCA

- Number of amendment steps

Their effect on the compression process and how we determine them is explained in the remainder of this section.

4.3.4.1 Number of principal components

The number of principal components used for an approximation of the data matrix for PPCA Offline method and the residual matrix for PPCA Online method is crucial with regard to the compression rates, as we can see in Equation (4.15). Since, we want to use the PPCA as a black box tool, besides a user given precision, we have to find a reliable way to determine the number of principal components. One can think about finding so-called spectral gaps, that can be found if the singular value is plotted against its index [79], see Figure 4.2 for an example. All singular vectors that are assigned to singular values left of the spectral gap are used as principal components. Since car crash simulations are highly nonlinear, see Remark 2.4, and the PCA is a linear model, a meaningful spectral gap can usually not be determined. For all existing methods, among others described in Section 4.1, we could not find any strategy to determine the number of principal components to achieve good compression rates under consideration of a required precision. Investigating Figure 4.2 we find big gaps between the first singular values and very small ones for higher indices. Unfortunately, the optimal number of principal components to gain the best possible compression rate with the PPCA method is usually neither indicated by a big spectral jump. We, therefore, developed an optimization process that estimates the possible compression rate on several sampling points to find the number of principal components that gives the best compression rate.

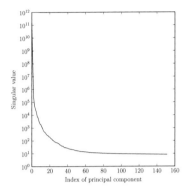

Figure 4.2: *Plot of singular values for the x coordinates of the driver's door from 3 Chrysler Silverados [98]. For this example, the combination from our PPCA Offline method with a Rice encoder leads to the best compression factor for 21 principal components. From 21 principal components to 22 is no significant jump which makes it hard to determine the optimal number of principal components based on the singular value's decay.*

We face the following trade-off to determine the optimal number of principal components with regard to the overall compressed size. When we use an insufficient number of principal components, the PCA prediction, see (4.13), results in high residuals and the residual matrix, which has the original dimension, is hard to compress, see (4.15). If we use more principal components than the optimum, the advantage for compressing the residual matrix is neglectable but the number of principal components as well as coefficients increases. This counteracts the idea of dimensionality reduction and increases the overhead. Therefore, we use a property which is valid for a big number of parts and state it as the following Remark.

Remark 4.11. *The function*

$$f : \mathbb{N} \to \mathbb{N} \ \textit{with} \ f : p \mapsto D(p)$$

that maps the number of used principal components to its overall compressed size is rapidly decreasing to a minimum and afterwards increasing slightly. Therefore, f is a unimodal discrete function, see [99].

Figure 4.3 shows as an example the compressed sizes for the driver's door for several numbers of used principal components. We apply a Fibonacci search [72] to determine the minimum of the discrete unimodal function f. For the Fibonacci search we determine a left boundary l, a right boundary r, a point in the middle $m = \left\lceil \frac{r-l}{2} \right\rceil$ and another point between the left and the middle point $k = \left\lfloor \frac{m-l}{2} \right\rfloor$. We evaluate the overall compressed size for applying m and k principal components, respectively. If $f(k) \leq f(m)$ the minimum has to lie in the range $[l, m]$. For $f(k) > f(m)$, the minimum has to lie in the interval $[m, r]$. Afterwards we select $[l, m]$ as new search interval and proceed.

In the case of 0 principal components, an alternative compression approach is used which

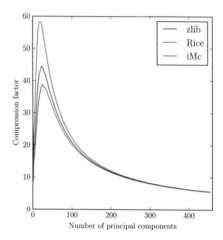

Figure 4.3: *Plot of compression factor for the x coordinates of the driver's door from 3 Chrysler Silverados [98] applying three different encoders on the residual. The overall compression size versus number of used principal components yields a convex-like function.*

exploits the dependency between adjacent time steps, see [94]. Therefore, the compression rate is usually significantly higher than an approach that uses only one principal component. In the optimization process, this value is initially ignored until the optimum is found using the principal components. This optimum is then compared with the compression rate in the case of 0 principal components. Thus, for each component, it is decided whether compression is performed using principal components or whether the time difference method from [94] is used, which has no advantage in the handling of sets of simulation results.

For performance reasons we do not optimize over the set of all possible dimensions $\mathcal{I} = \{1, ..., \min\{n, t\}\}$ for n number of data points and t number of points in time. First, we restrict the interval to $\hat{\mathcal{I}} = \{1, ..., \lfloor \frac{1}{4} \min\{n, t\} \rfloor\}$, since using more than $\frac{1}{4}$ of all principal components, we have almost no dimensionality reduction. Second, we cluster 1 up to 5 numbers to buckets and determine the compression rate for one element of these buckets. The sampling points can for example be $\tilde{\mathcal{I}} = \{1, 2, 3, 5, 7, 11, 15, ..., \lfloor \frac{1}{4} \min\{n, t\} \rfloor\}$. Since one property of f is, that the changes in compression rate are much higher for a small number of principal components compared the case of a big p, we sample in more detail for small number of principal components. With this strategy, even if we do not find the exact optimum, the overall compression rate is close to the optimum.

For the Fibonacci search we have to evaluate the overall compressed size for several sampling points. We need to compress the mean values, the principal components, the coefficients, and the residual, see (4.11), (4.12), and (4.13). For this we apply an entropy encoder, see Section 3.3.3.1, where we can directly determine the compressed size based on the statistics of the data stream. We developed the iMc encoder, see Section 5.1 to meet this requirement. Applying the iMc encoder, we can save the arithmetic encoding since it is sufficient to build up the statistics and calculate the compressed size via the entropy of the data stream.

Although we cannot guarantee that f is unimodal for all parts we achieve good overall compression rates because the trend on a global scale is the same. Due to run time considerations, we also accept local optimums.

Other techniques to find the optimal number of principal components such as Akaike's information theoretic criterion or minimum description length criterion have not yet been applied, cf. [20].

4.3.4.2   Internal precision

In this section, we focus on the internal precision for mean values, principal components, and coefficients, that is specified by the quantum q, see e.g. (4.11) and (4.12).

We quantize the PCA components for two reasons. On the one hand we achieve better compression rates if we apply a lossy compression on these components. On the other hand, when we have quantized values, we can use integer arithmetic. While floating point arithmetic is not uniform for all machines, operating systems and cores, it is for integer arithmetic. Therefore, all multiplications and summations for the reconstruction is based on integer data. Only the dequantization as the last step of decompression and one additional summation, see (4.22), can result in minimal differences on different machines. Therefore, we have to find one uniform precision for mean values, principal components as well as coefficients to perform the reconstruction without an intermediate dequantization. One drawback of using a uniform quantization is, that principal components and coefficients, which are associated with a small singular value, would be stored as precise as those who are associated with big singular values. Therefore, before applying a uniform quantization, we split the singular values and scale the coefficients as well as the principal components with the square root of the associated singular value, see (4.12). Thus, we have an adaptive quantization that adapts its precision based on the importance of principal components and the coefficients to the given data stream matrix. Since we have to scale the principal components in the PPCA Online method to ensure convergence, see Remark 4.7, the effect on the overall run time is neglectable. The mean values are not effected as they are not scaled. From Theorem 4.6 and its proof, we know that the effect of applying a uniform quantization on the mean values, generates an error of not more than $\frac{q}{2}$ for every element of the data stream.

It is a common problem for lossy compression to find a meaningful precision. We face the following conflict of objectives: If the internal precision for the mean values, principal components, and coefficients is too low, we will have an insufficient prediction and the residual will be hard to compress. If the internal precision is too high, the data sets themselves are hard to compress and prevent a good compression rate, and the optimization process for the number of principal components would determine only a small amount of principal components. Therefore the PPCA Online method would not provide a sufficient base for learning.

Another important criterion is that the internal precision effects the functionality of the amendment steps of the PPCA Online method, see Section 4.3.3, which is investigated in the following. Since the amendment steps are an application of the Richardson iteration, see Remark 4.8, their convergence criteria are the same like for splitting methods.

Theorem 4.12 ([48]). *Suppose $b \in \mathbb{R}^P$ and $D = M - N \in \mathbb{R}^{P \times P}$ is nonsingular. Moreover* *let M be nonsingular. The spectral radius of $M^{-1}N$ satisfies the inequality $\rho(M^{-1}N) < 1$ if and only if the iterates $x^{(k)}$ defined by $Mx^{(k+1)} = Nx^{(k)} + b$ converge to $x = D^{-1}b$ for any starting vector $x^{(0)}$.*

For the Richardson iteration, we have $M = I$ and $N = I - D$ with Gram matrix $D = \hat{V}^T\hat{V}$. For convergence, we have to limit the maximal absolute eigenvalue of $N := I - D$ to less than 1. The Gram matrix D can be written as

$$D = \hat{V}^T\hat{V} = (d_{ij})_{i,j=1,\dots,P} \in \mathbb{R}^{P \times P} \text{ with } d_{ij} = \langle v_i^{\lfloor q \rfloor} | v_j^{\lfloor q \rfloor} \rangle . \tag{4.34}$$

With matrix D being symmetric for all eigenvalues $\lambda \in \text{spec}(D)$ it holds that $\lambda \in \mathbb{R}$. We summarize the meaning of matrix D in the following remark.

Remark 4.13. *Each nonzero off-diagonal element of D indicates missing orthogonality of the lossy compressed principal components. Every diagonal entry D_{ii} unequal to one shows that the i-th principal component is not normalized any more. Since we perform a renormalization, see (4.24), the diagonal entries of D will be one.*

Next, we determine a sufficient precision to guarantee that the eigenvalues of the iteration matrix N do not exceed 1. Due to the dimensionality reduction we will have $n - P$ times an eigenvalue of 1, see Lemma A.14. For $n > P$, there is no internal precision that results in the necessary condition for convergence: $\|K\|_2 < 1$, see Theorem 4.12.

Theorem 4.14. *Let $N = I - D \in \mathbb{R}^{P \times P}$ with $D = \hat{V}^T\hat{V}$, $\hat{V} \in \mathbb{R}^{n \times P}$ from (4.24), quantum $q \in \mathbb{R}_{>0}$ and $P \leq n$ but $P > 1$. Let the singular values $\sigma_1, \dots, \sigma_P$ be given in descending order. We define*

$$Q_{ji} := (2P - 1)\frac{\sqrt{\sigma_j} + \sqrt{\sigma_i}}{\sqrt{n}} - \sqrt{\frac{(2P-1)^2}{n}(\sqrt{\sigma_j} + \sqrt{\sigma_i})^2 - \frac{4}{n}\sqrt{\sigma_i\sigma_j}}, \ \forall i, j = 1, \dots, P.$$

For all eigenvalues $\lambda \in \text{spec}(N)$ it holds that $\|\lambda\| < 1$, if

$$q < Q_{ji} \ \forall i, j = 1, \dots, P \tag{4.35}$$

together with

$$q < \frac{2\sqrt{\sigma_P}}{\sqrt{n}} \tag{4.36}$$

and therefore convergence of the Richardson iteration for (4.26).

For $P = 1$, we do not have an issue due to missing orthogonality, since we just have one principal component. For the proof of Theorem 4.14 we need the Gerschgorin Circle Theorem.

Theorem 4.15 ([48]). *Let $A = B + C \in \mathbb{C}^{P \times P}$ with $B = \text{diag}(d_1, \dots, d_P)$ and $C_{ii} = 0$ for all $i = 1, \dots, P$. For all $\lambda \in \text{spec}(A)$, it holds that*

$$\lambda \in \bigcup_{i=1}^{P} B_i \text{ with } B_i = \{z \in \mathbb{C} : |z - d_i| \leq \sum_{j=1}^{P} |C_{ij}|\}. \tag{4.37}$$

It follows the proof of Theorem 4.14.

PROOF: We investigate the off-diagonal elements of matrix D to approximate its spectrum via Gerschgorin Circle Theorem. Let $i \in \{1, ..., P\}$ and $\hat{v}_i \in \mathbb{R}^n$ be the i-th column vector of $\hat{V}$ and $v_i \in \mathbb{R}^n$, $n \in \mathbb{N}\{0\}$ the vector before quantization and scaling for all $i = 1, ..., P$. We define

$$w_i = \frac{q}{\sqrt{\sigma_i}} (v_i) \left\lfloor \frac{q}{\sqrt{\sigma_i}} \right\rceil$$

and $\hat{e}_i = \frac{w_i}{\|w_i\|_2} - v_i$. The vector $\hat{v}_i$ can be expressed as the renormalization of w_i that is a reconstruction of v_i after quantization with quantum $\omega_i = \frac{q}{\sqrt{\sigma_i}}$:

$$\frac{w_i}{\|w_i\|_2} = \frac{\frac{q}{\sqrt{\sigma_i}} \left\lfloor \frac{\sqrt{\sigma_i} v_i}{q} + \frac{1}{2} \right\rfloor}{\sum_{j=1}^{n} \frac{q}{\sqrt{\sigma_i}} \left\lfloor \frac{\sqrt{\sigma_i} (v_i)_j}{q} + \frac{1}{2} \right\rfloor}$$

$$= \frac{q \left\lfloor \frac{\sqrt{\sigma_i} v_i}{q} + \frac{1}{2} \right\rfloor}{\sum_{j=1}^{n} q \left\lfloor \frac{\sqrt{\sigma_i} (v_i)_j}{q} + \frac{1}{2} \right\rfloor}$$

$$= \frac{\tilde{v}_i}{\|\tilde{v}_i\|_2}$$

$$= \hat{v}_i.$$

Let $e_i = w_i - v_i$ be the error of the absolute quantization of the normalized vector v_i. The vectors $\hat{v}_i$ are normalized, i.e. $\|\hat{v}_i\|_2 = 1$ and therefore all diagonal entries of D are equal to one. The vectors v_i are orthonormal, i.e. $\|v_i\|_2 = 1$ and $\langle v_i | v_j \rangle = 0$ for all $i, j = 1, ..., P$. When using a uniform quantization on a scalar the maximal error of reconstruction is $\frac{\omega_i}{2} = \frac{q}{2\sqrt{\sigma_i}}$, see (3.28). Considering the equivalence of norms, we have

$$\|e_i\|_2 \leq \sqrt{n} \|e_i\|_\infty \leq \frac{q\sqrt{n}}{2\sqrt{\sigma_i}} \forall i \in \{1, ..., P\}. \tag{4.38}$$

Due to the renormalization, the quantization error of w_i and the reconstruction error of $\hat{v}_i$ differ. From $q < \frac{2\sqrt{\sigma_i}}{\sqrt{n}}$, we get an upper bound for the quantization error: $\|e_i\|_2 < 1$. Applying the reverse triangle inequality [2], we get:

$$\|\hat{e}_i\|_2 = \|\hat{v}_i - v_i\|_2$$

$$= \left\| \frac{v_i + e_i}{\|v_i + e_i\|_2} - v_i \right\|_2$$

$$= \frac{\|(1 - \|v_i + e_i\|_2)v_i + e_i\|_2}{\|v_i + e_i\|_2}$$

$$\leq \frac{|1 - \|v_i + e_i\|_2| \overbrace{\|v_i\|_2}^{=1} + \|e_i\|_2}{\|v_i + e_i\|_2}$$

$$\leq \frac{1 - (\|v_i\|_2 - \|e_i\|_2) + \|e_i\|_2}{\|v_i\|_2 - \|e_i\|_2}$$

$$= \frac{2\|e_i\|_2}{1 - \|e_i\|_2}. \tag{4.39}$$

We investigate the off-diagonal elements of matrix N:

$$|\langle \hat{v}_i | \hat{v}_j \rangle|$$

$$= |\langle v_i + \hat{e}_i | v_j + \hat{e}_j \rangle|$$

$$= \left| \underbrace{\langle v_i | v_j \rangle}_{=0} + \langle \hat{e}_i | v_j \rangle + \langle v_i | \hat{e}_j \rangle + \langle \hat{e}_i | \hat{e}_j \rangle \right|$$

$$\leq \|\hat{e}_i\|_2 \underbrace{\|v_j\|_2}_{=1} + \|\hat{e}_j\|_2 \underbrace{\|v_i\|_2}_{=1} + \|\hat{e}_i\|_2 \|\hat{e}_j\|_2$$

$$= \|\hat{e}_i\|_2 + \|\hat{e}_j\|_2 + \|\hat{e}_i\|_2 \|\hat{e}_j\|_2$$

$$\overset{(4.39)}{\leq} \frac{2\|e_i\|_2}{1 - \|e_i\|_2} + \frac{2\|e_j\|_2}{1 - \|e_j\|_2} + \frac{2\|e_i\|_2}{1 - \|e_i\|_2}\frac{2\|e_j\|_2}{1 - \|e_j\|_2}$$

$$= \frac{2(\|e_i\|_2 + \|e_j\|_2)}{(1 - \|e_i\|_2)(1 - \|e_j\|_2)}$$

$$\overset{(4.38)}{\leq} \frac{(\sqrt{\sigma_j} + \sqrt{\sigma_i})\sqrt{nq}}{(\sqrt{\sigma_i} - \frac{\sqrt{nq}}{2})(\sqrt{\sigma_j} - \frac{\sqrt{nq}}{2})}$$

$$\overset{(4.35),(4.36)}{<} \frac{Q_{ij}\sqrt{n}(\sqrt{\sigma_i} + \sqrt{\sigma_j})}{\sqrt{\sigma_i}\sqrt{\sigma_j} - (\sqrt{\sigma_i} + \sqrt{\sigma_j})\frac{\sqrt{n}}{2}Q_{ij} + \frac{n}{4}Q_{ij}^2} \qquad (4.40)$$

$$\overset{\sigma_i>0,\sigma_j>0,n>0}{\leq} \frac{(2P-1)(\sqrt{\sigma_i}+\sqrt{\sigma_j})^2-(\sqrt{\sigma_i}+\sqrt{\sigma_j})2\frac{\sqrt{n}}{2}\sqrt{\frac{(2P-1)^2}{n}(\sqrt{\sigma_i}+\sqrt{\sigma_j})^2-\frac{4}{n}\sqrt{\sigma_i}\sqrt{\sigma_j}}}{(-\frac{2P-1}{2}+2(\frac{2P-1}{2})^2)(\sqrt{\sigma_i}+\sqrt{\sigma_j})^2-(P-1)(\sqrt{\sigma_i}+\sqrt{\sigma_j})\sqrt{(2P-1)^2(\sqrt{\sigma_i}+\sqrt{\sigma_j})^2-4\sqrt{\sigma_i}\sqrt{\sigma_j}}}$$

$$\overset{(4.35),\sigma_i>0,\sigma_j>0,n>0}{=} \frac{1}{P-1} \qquad (4.41)$$

Since all diagonal entries of D are equal to 1, we can just select the biggest ball B_i of (4.37) to find the eigenvalue with the biggest distance to 1 with the Gerschgorin Circle Theorem. We have for all $i, k \in \{1, ..., P\}$ with $\lambda_k \in \text{spec}(D)$

$$|\lambda_k - 1| = |\lambda_k - \langle \hat{v}_i | \hat{v}_i \rangle|$$

$$\leq \max_{i \in \{1,...,P\}} \left\{ \sum_{j \in \{1,...,P\}\setminus\{i\}} |\langle \hat{v}_i | \hat{v}_j \rangle| \right\}$$

$$\overset{(4.41)}{<} (P-1)\frac{1}{P-1}$$

$$= 1 \qquad (4.42)$$

Since $D \in \mathbb{R}^{P \times P}$ is symmetric it has real eigenvalues. Therefore, each eigenvalue of D lies in the interval of $(0, 2)$. The Richardson iteration has a iteration matrix $N = (I - D)$. Since all eigenvectors of D are also eigenvectors of $N = (I - D)$, it holds that

$$\text{spec}(N) = \{1 - \lambda : \lambda \in \text{spec}(D)\}.$$

Therefore, all eigenvalues of matrix N lie in the interval $(-1, 1)$. Together with Theorem 4.12, the statement follows. $\qquad \square$

The singular values are sorted in a descending order leading to the following approximation.

Corollary 4.16. *Let* $N = I - D \in \mathbb{R}^{P \times P}$ *with* $D = \hat{V}^T \hat{V}$, $\hat{V} \in \mathbb{R}^{n \times P}$ *from (4.24), quantum* $q \in \mathbb{R}_{>0}$ *and* $P \leq n$ *but* $P > 1$. *Satisfying*

$$q < \tilde{q} := \frac{2\sqrt{\sigma_P}}{\sqrt{n}} \left((2P - 1) - 2\sqrt{P(P-1)} \right) \tag{4.43}$$

it also holds that $q < Q_{ij}$ *for all* $i, j \in \{1, ..., P\}$ *with* Q_{ij} *from Equation (4.35).*

PROOF: We define the function $f : \mathbb{R}^+ \times \mathbb{R}^+ \to \mathbb{R}^+$ with $A = 2P - 1$ and $P \in \mathbb{N} \wedge P > 1$ as

$$f(x, y) = A\frac{x + y}{\sqrt{n}} - \sqrt{\frac{A^2}{n}(x+y)^2 - \frac{4}{n}xy}.$$

The function f is continuously differentiable since it is a combination of continuously differentiable functions.

$$\frac{\delta}{\delta x} f(x, y) = \frac{1}{\sqrt{n}} \left(A - \frac{A^2(x+y) - 2y}{\sqrt{A^2(x+y)^2 - 4xy}} \right).$$

$$f(x, y) = A\frac{x + y}{\sqrt{n}} - \sqrt{\frac{A^2}{n}(x+y)^2 - \frac{4}{n}xy} = f(y, x).$$

For $x, y \in \mathbb{R}^+$ it holds

$$0 < y$$
$$\Leftrightarrow 0 > -y^2(A^2 - 1)$$
$$\Leftrightarrow 0 > -4y^2 A^2 + 4y^2$$
$$\Leftrightarrow A^4(x^2 + 2xy + y^2) - 4xy A^2 > A^4(x^2 + 2xy + y^2) - 4y A^2(x+y) + 4y^2$$
$$\Leftrightarrow A^2(A^2(x+y)^2 - 4xy) > (A^2(x+y) - 2y)^2$$
$$\Leftrightarrow \left| A\sqrt{A^2(x+y)^2 - 4xy} \right| > |A^2(x+y) - 2y|$$
$$\Leftrightarrow A\underbrace{\sqrt{A^2(x+y)^2 - 4xy}}_{>0} > A^2(x+y) - 2y$$
$$\Leftrightarrow A > \frac{A^2(x+y) - 2y}{\sqrt{A^2(x+y)^2 - 4xy}}$$
$$\Leftrightarrow 0 < \frac{1}{\sqrt{n}} \left(A - \frac{A^2(x+y) - 2y}{\sqrt{A^2(x+y)^2 - 4xy}} \right)$$
$$\Leftrightarrow 0 < \frac{\delta}{\delta x} f(x, y)$$

The function f is symmetric in its two variables. Therefore it also holds that $0 < \frac{\delta}{\delta y} f(x, y)$. Therefore for functions $f_y : \mathbb{R}^+ \to \mathbb{R}^+$ with $f_y : x \mapsto f(x, y)$ and $f_x : \mathbb{R}^+ \to \mathbb{R}^+$ with $f_x : y \mapsto f(x, y)$. Since the derivatives are positive in both x and y direction, the functions are strictly monotone. Let all $x, y, h, k \in \mathbb{R}^+$.

$$f(x, y) = f_y(x) < f_y(x + h) = f(x + h, y) = f_{x+h}(y) < f_{x+h}(y + k) = f(x + h, y + k)$$

The smaller the input is, the small will be the function f. Since all singular values $\sigma_1, ..., \sigma_P > 0$ are sorted in an descending order, the function is minimal for $x = y = \sigma_P$.

$\square$

For the data stream of the driver's door from a set of three Chrysler Silverados, the PPCA Offline method combined with iMc encoding is optimal applying $P = 18$ principal components. The 18th singular value is $\sigma_{18} = 259.94$. The door has 8375 nodes, which results in a required quantum of $\min\{0.35235, 0.02877\} = 0.02877$ guaranteeing the convergence of the Richardson iteration, see Theorem 4.14 and Corollary 4.16.

Unfortunately, the bound on quantum q of (4.35) is usually not sharp. Actually, demands on precision are high, since we assume that all quantization errors accumulate, which is rather rare. Our evaluation shows that a quantum of $q = 2^{-8}$ achieves good compression ratios and results in convergence of the amendment steps for our empirical investigations, see Chapter 6. Since we store the residual of the prediction by the PPCA components, even in the case of no convergence the PPCA approach will not fail. Therefore, we can guarantee the user defined precision Q.

4.3.4.3   Minimal size of data streams processed by PPCA

The necessity to define a minimal number of data points n as well as points in time t to apply the PPCA is obvious by the following consideration. The PPCA Offline method generates an overhead when compressing $n \cdot t$ elements of a data stream matrix $A \in \mathbb{R}^{n \times t}$ for mean values, principal components and coefficients by $n + n \cdot P + P \cdot t$, when we use P principal components. The idea of PPCA methods is to generate an easy to compress residual matrix by hard to compress elements like principal components and coefficients. If n and t are small, the number of principal components P is usually big compared to n and t. Therefore, the advantage of dimensionality reduction is usually not visible in such cases.

A simulation result file often contains many small parts with less than 40 nodes, see Remark 2.4. Due to the overhead of PPCA, a part of this size can normally be compressed better by prediction in time and space. We, therefore, do not apply the PPCA on these parts, but use an alternative compression method or we merge several parts to achieve a certain size. Compressing merged parts might be problematic. First, we can only add parts that have a similar behavior. Adding a rigid body part, which cannot be deformed by the crash with a deformable part in the crash area, brings no improvements compared to state-of-the-art compression methods for one simulation result. This problem can be handled by applying clustering methods. Secondly, the merging of parts only makes sense, if we know that these parts exist in almost all simulation results. If they do not, we have to add zeros for non existing parts and generate a certain overhead. To avoid such problems we set an internal threshold to only process parts with at least 40 data points with our PPCA methods. Moreover, for the PPCA Offline method we demand at least 20 points in time.

4.3.4.4 Number of amendment steps

We apply amendment steps in the compression phase of the PPCA Online method to tackle the missing orthogonality of the quantized and dequantized principal components from the database, see Section 4.3.3. Especially, if the components are linearly dependent, the reconstruction error will also have components pointing in this direction. This is due to the calculation of coefficients by multiplication of the quantized and dequantized principal components with the centralized matrix. Therefore, we update the coefficients to eliminate multiple influences facing in the same direction.

The application of amendment steps is necessary only during the compression phase, since we only update the coefficients for the newly added simulation model. Since we have an asymmetric compression and all calculations are based on the dimensions $n \times P$ and $P \times t$, where n is the number of nodes, P the number of used principal components, and t the number of time steps, this is a fast procedure.

In Section 4.3.4.2, we investigate the spectrum of Gram matrix $D = \hat{V}^T \hat{V}$ to get a sufficient quantum q to ensure convergence of the amendment steps if they are interpreted as the Richardson iteration on the normal equation of dimension P, see Remark 4.8. But the amendment steps can also be represented b an iteration matrix $K = (I - \hat{V}\hat{V}^T) \in \mathbb{R}^{n \times n}$ that is multiplied with the residual, see (4.28). In this case, the errors due to nonorthogonality become smaller for each amendment step if q is small enough, see Theorem 4.14, but the errors due to dimensionality reduction remain, see Lemma A.14 and Equation (4.30). To avoid the accumulation of errors due to finite machine precision we limit the number of amendment steps with respect to the maximal absolute eigenvalue λ of iteration matrix $N = I - D$, see Theorem 4.14. If we have a required residual reduction by factor of $\delta \in (0, 1)$, then $a \geq \frac{\ln \delta}{\ln \lambda}$ amendment steps are sufficient. Next, we have two options to determine the maximal absolute eigenvalue of N and a sufficient number of amendment steps:

- determine an a priori bound based on the limits on the spectrum of D for a given quantum q, see Equation (4.42),

- determine D and find eigenvalue bounds based on Gerschgorin Circle Theorem, see Theorem 4.15.

The first approach seems not promising since the estimations are not exact enough. Therefore, the second approach is preferable since it adapts its behavior to the given data stream. A third option is to predefine a fixed number of amendment steps. Empirical tests show that 10 amendment steps are sufficient for all investigated models, see Section 6.3.

4.3.5 Segmentation into parts

In this section we analyze what type of parts we can handle with PPCA.

If we want to describe the crash of a car for all 1 up to 10 million nodes in one sweep principal components have to match this dimension. With the number of time steps being usually limit by 60 for one simulation model only few principal components are possible. Since many parts sow a different behavior during a crash test and have to express all different behaviors with few components, this approach would not be efficient. Therefore, we decided to operate only on parts.

There are several ways to segment a car crash model into smaller components:

- parts defined by input deck

- parts defined by octree decomposition of a bounding box of the model

- hybrid methods, which basically use the part definition by input decks but attach small parts to bigger parts and decompose big parts in smaller ones.

The obvious choice is to take the part definition by the input deck as we can benefit from recurring parts. Design changes often effect only a few parts and for an offline method with several simulation results or an online method, we profit from exploiting the knowledge from already processed parts. One disadvantage is that usual model comprises several hundred parts with less than 40 nodes and a dimensionality reduction method on such small components cannot easily be applied. Moreover, we accept a certain overhead by duplicating nodes that are part of more than one part in the input deck, see Section 2.2. To decompose a bounding box of the initial model and refines every box until a threshold of a maximum number of nodes is undershot an octree can be applied. This approach also allows to bind nodes from parts with different behavior such as a rigid body element and a shell element. Especially, in the detailed resolved engine compartment, it is problematic to find meaningful principal components for both behaviors simultaneously. A hybrid method combines both approaches. Parts that are too big to be handled simultaneously can be decomposed, e.g. by an octree method or the application of methods by [117]. Moreover, small parts are connected to each other or to bigger parts. Therefore, we can use information about connections to glue these small parts and get bigger parts. Unfortunately, the reuse of glued parts can be limited, see Section 4.3.4.3. The investigation of gluing and decomposing components will be part of future work.

Furthermore, a property of car crash simulation result files, see Remark 2.4, is that they contain auto generated parts that contain the same nodes but have a different mesh. These cases can be handled in two ways. The first approach is to identify parts with different topologies of the mesh as different parts. The second one is to store two lists of connectivities for such parts, make an assignment of which topology was used for the simulation, and consider the time dependent data as it would be from the same part. For the PCA methods itself the topology is not crucial as long as the number of nodes and the locations of these nodes coincide. Therefore, we decided to compress the node values for such parts with different connectivities together.

Conclusion:

In this chapter, we have introduced a new way of using PCA, i.e. PPCA, for data compression of simulation result sets, that uses the PCA as prediction method and perfectly fits into a SDMS, see Remark 2.5. Especially, in the case of tight point wise bounds on the reconstruction error, this method can efficiently be applied, see Chapter 6. The aim is to generate an overhead by storing PCA elements like coefficients and principal components, which are hard to compress successfully only for a small internal dimensions, and achieve a residual that has the full dimension to be efficiently compressed with our iMc encoder as detailed in Chapter 5. A main difference to all PCA applications in academic literature, see Section 4.1, is that we determine the intrinsic dimension almost optimal with a black box method, see Section 4.3.4.1. For this purpose, we estimate the obtained compression rates that can be processed by the iMc encoder by only setting up the statistics and calculating the entropy. Therefore, we can omit the encoding step in the optimization process.

When all data sets are given simultaneously, we apply the PPCA Offline simultaneously simulation results at once and exploit similarities between the simulation results. In a real world scenario the data sets are available immediately after simulation and therefore sequential. To not keep data uncompressed we developed the PPCA Online method, that reuses information of the initial PPCA Offline method and all PPCA Online steps afterwards to compress a new simulation result. An important property is that we do not alter the data for previously compressed simulation results but achieve a learning process by adding new principal components. We overcome the issues of missing orthogonality of the lossy compressed principal components by fast amendment steps, see Section 4.3.3 and 4.3.4.4. Those are an iterative application of the Richardson iteration on the normal equation to find the optimal coefficients for embedding, see Remark 4.8. We find an a priori bound on the internal precision to be applied on the PCA elements to guarantee the convergence of the amendment steps. The a priori bound is not sharp for real world cases. We, therefore, choose internal precision that is less precise to gain a better compression performance. In Chapter 6, we see that our internal precision is sufficient. Since we store the residual anyway we can guarantee the reconstruction with a user defined precision, although the amendment steps would not converge.

Furthermore, we introduced a bound on the residual effected by the internal precision, the number of used principal components, and the singular values of the data stream matrix, see Theorem 4.6. Although, the estimation is not sharp, we see that the elements of the residual matrix are limited, which is a good property for our iMc encoder introduced in Chapter 5.

Chapter 5

Induced Markov chains for data compression

In this chapter, we introduce the induced Markov chains (iMc). We use the iMc to model the statistical relations of graph-based data, which we apply as side information in a statistical encoder. For a short description of an iMc application we refer to [96].

Throughout this thesis, our strategy is to describe the dependencies of an input stream statistically and exploiting the dependencies between values of a stream. The proposed encoder exploits the relations resulting from the topology of the graph. In the context of crash test simulations the topology is compressed lossless, see Remark 2.4. Moreover, the topology is used in several state-of-the-art prediction methods such as Lorenzo predictor and tree differences, see Section 3.3.2. Apart from a few approaches in the context of graphical models for graph-based data, there is no encoding scheme available that exploits the topological information in the encoding phase. If the topology is exploited, it will be used in the prediction phase. Empirical investigations, see e.g. [96] show that topological dependencies exists after applying a topological prediction method. We model such dependencies with the iMc in contrast to common encoding methods used for compression of simulation results. With this approach we aim for better compression rates.

The induced Markov chains aim at developing a new graphical model that does not need an individual statistic per node but per node value. Still, the used statistics should reflect the topological relation between the nodes. The iMc is implemented as a black box encoder and has to work properly on well-predicted as well as on not accurately predicted data sets. We focus on entropy encoders not only, but also because of the optimization process in the Online and Offline stage of PPCA. After having determined a data specific statistic we can approximate the compression ratio based on the entropy, if we use a suitable encoder scheme that achieves at least approximately the limit of compression. We can avoid computationally expensive bit operations for finding the optimum. Having found the optimum we start the actual encoding using the Arithmetic Encoder introduced in Section 3.3.3.1.

In Section 5.1, we start with an application of our iMc encoder. We proceed with an investigation of the time complexity of the iMc encoder in Section 5.2 and compare it with those of Markov Random Fields, see Section 3.3.3.7. In Section 5.3, we introduce an adaptive variant of the iMc encoder. This approach exploits that we store the number of occurrences of each symbol by eliminating the already visited node values from the statistics and only keep the ones that will occur in the future. For the mathematical background of our iMc encoder we refer to Section 5.4. In Section 5.5 we determine bounds on how good the model of induced Markov chains fits to a given data set.

5.1 The induced Markov chain encoder

The iMc encoder is a practical implementation of an arithmetic encoder applying statistics that we will categorized as induced Markov chains, see Section 5.4. The mathematical background of induced Markov chains is important to prove certain properties of the iMc encoder but it is not required to understand the implementation of the iMc encoder. We, therefore, decided to move the practical implementation to the beginning. The workflow of the iMc encoder is shown in Figure 1.3.

The modeling of graph-based data as an iMc can be used in an arithmetic encoder that exploits connectivities of the graph and statistics as side information. A possible implementation is to establish the graph, then define a spanning tree. Determining a tree of the graph allows to consider the tree as a directed graph with the direction of all edges from the root to the leafs. Based on the directed tree we can determine the transition probabilities that are the relative frequencies of transitions from one value to another one. Afterwards, we use the tree as a directed graph and proceed from the root to the leafs and apply the transition probabilities as distributions in an arithmetic encoder.

As for the iMc encoder the root node value has to be known for reconstruction we store them uncompressed as side information in the case of big parts with only a few time steps. For many time steps and small parts we apply a prediction method over time in combination with an LZ encoder, e.g. the zlib [111]. To define our iMc encoder we have to define empirical probabilities and transition probabilities.

Definition 5.1 (empirical probabilities). *Let $G = (V, E)$ be a graph, $\mathcal{A} = \{a_1, ..., a_m\}$ an alphabet and $v \in \mathbb{Z}^n$ a vector with $n = \#V$ containing one integer for each node. Moreover, let a function $P : \mathcal{A} \to \mathbb{Q} \cap [0, 1]$ with*

$$P(a_k) := \frac{\#\{(i) \in V | v_i = a_k\}}{\#V} \tag{5.1}$$

that represents the relative frequencies of node values. We call all elements of $P_k := P(a_k)$ for $k, l \in \mathcal{A}$ empirical probabilities.

Definition 5.2 (transition probabilities, transition frequencies). *Let $G = (V, E)$ be a graph, $\mathcal{A} = \{a_1, ..., a_m\}$ an alphabet and $v \in \mathbb{Z}^n$ a vector containing one integer for each node. Moreover, let a function $P_t : \mathcal{A} \times \mathcal{A} \to \mathbb{Q} \cap [0, 1]$ with*

$$P_t(a_k, a_l) := \frac{\#\{(i, j) \in E | v_i = a_k \wedge v_j = a_l\}}{\#\{(i, \cdot) \in E | v_i = a_k\}} \tag{5.2}$$

that represents the relative frequencies of transitions from one node value to a neighbor's value be given. We call all elements of $P_{kl} := P_t(a_k, a_l)$ for $k, l \in \mathcal{A}$ transition probabilities. Moreover, we call the counter of $P_t(a_k, a_l)$ transition frequencies for graph G and node vector $v \in \mathbb{Z}^n$ and write

$$P_G^F := \#\{(i, j) \in E | v_i = a_k \wedge v_j = a_l\}. \tag{5.3}$$

An example for a graph with integer node values can be found in Figure 5.1 that concurrently is Model Problem 1.

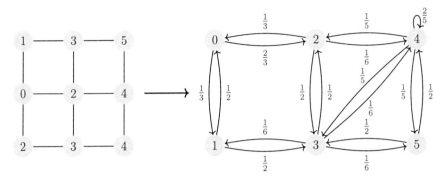

Figure 5.1: Mesh of Model Problem 1 with node values and the induced Markov chain with transition probabilities.

Model problem 1.

We assume that an undirected graph $G = (V, E)$ with n nodes and e edges together with an alphabet $\mathcal{A} = \{a_1, ..., a_m\}$ that contains all node values is given.

Remark 5.3. *The idea of the iMc encoder is:*

- *use the knowledge about the grid for encoding.*

- *transform a topological relation into a value-based one.*

- *define transition probabilities between occurring values that are dependent on the topology of the graph.*

- *the transition probabilities along with a root node value can be used as side information for an encoder.*

- *the transition probabilities along with the empirical probabilities of node values creates a Markov chain.*

For non-root elements of the tree, we know the node value of their predecessor and can apply the transition probabilities specified for the given tree, cf. Figure 5.5. This results in an encoding scheme whereby the applied distribution in each step depends on the current node value and can change from one node to another. Changing distributions require an encoding scheme that can be adapted in every encoding as well as decoding step. We apply the Arithmetic encoder, see Section 3.3.3.2 and [93, 115, 116].

There are several possibilities of finding a spanning tree of a connected graph such as breadth-first search and depth-first search [24]. To find a spanning tree we can either use a deterministic strategy with no need to store the structure as it can be reproduced during decompression. Or we can store the tree instantly if it has to be replicated, e. g in every time step and for different variables.

Remark 5.4. *We use a sparse matrix format for the description of the transition proba-*
bilities, since we observe that in the case of a big alphabet with several thousand letters the
number of nonzero entries is usually linear in the size of the alphabet. Whereas the size
of the matrix is equal to the size of the alphabet squared. When the data is successfully
decorrelated we usually have an alphabet with less than 500 letters and the overhead of
using the CSR format for an almost dense matrix of this size is small if it exists at all.

If the mesh and the node values are available, the main steps of the encoding algorithm
are:

1. construct the adjacency matrix in compressed sparse row format, see Figure 5.2

2. find a spanning tree by depth first search method, see Figure 5.3

3. build the adjacency matrix of the tree, see Figure 5.4

4. store root values and optionally apply tree and time differences, see Figure 5.5

5. determine transition frequencies, see Figure 5.6

6. cumulate the transition frequencies (for arithmetic encoding), see Equations (5.5)
   and (5.4)

7. determine a map matrix (for faster access due to the CSR format), see Equation
   (5.7)

8. execute the arithmetic coding, see Figure 5.7 and 5.8

9. encoding of the transition frequency matrix.

In the first step, we construct the adjacency matrix of graph G. We use a sparse matrix
format since our main application area are graphs based on Finite Element grids with the
number of neighbors being small compared to the number of nodes, see Figure 2.4. The
sparse matrix representation is beneficial in the sense of run time and memory consumption compared to a dense matrix. As we are only interested in the topology of the graph,
we omit the value vector of the CSR representation of the matrix, since for a standard
adjacency matrix every nonzero value would be one.

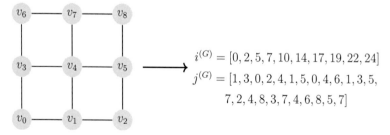

$$i^{(G)} = [0, 2, 5, 7, 10, 14, 17, 19, 22, 24]$$
$$j^{(G)} = [1, 3, 0, 2, 4, 1, 5, 0, 4, 6, 1, 3, 5,$$
$$7, 2, 4, 8, 3, 7, 4, 6, 8, 5, 7]$$

Figure 5.2: Construct the adjacency matrix of G in compressed sparse row (CSR) format
[113]. Since we are only interested in the location of nonzero values we omit the value
vector of the matrix.

In the next step we create a spanning tree T based on the graph G by an breadth first search technique, see [24]. In contrast to the undirected graph G the tree has a direction from the root node to its leafs.

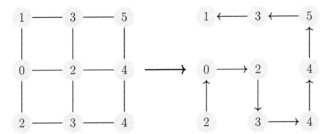

Figure 5.3: *Application of the depth first search algorithm of boost on the adjacency matrix of graph G to find a spanning tree T.*

Same as for the graph G in the first step we apply the tree via its CSR representation. We construct the adjacency matrix of T. As we are only interested in the topology we again omit the value vector in the CSR matrix.

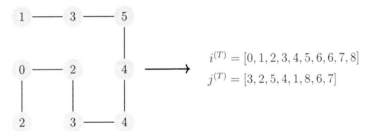

$$i^{(T)} = [0, 1, 2, 3, 4, 5, 6, 6, 7, 8]$$
$$j^{(T)} = [3, 2, 5, 4, 1, 8, 6, 7]$$

Figure 5.4: *Construct the adjacency matrix of tree T in CSR format [113]. As we are only interested in the location of nonzero values we again omit the value vector of the matrix.*

The transition probabilities as well as the root node value are used as side information while encoding and are also needed for the decoding. The side information and the values that we encode are listed in the following figure.

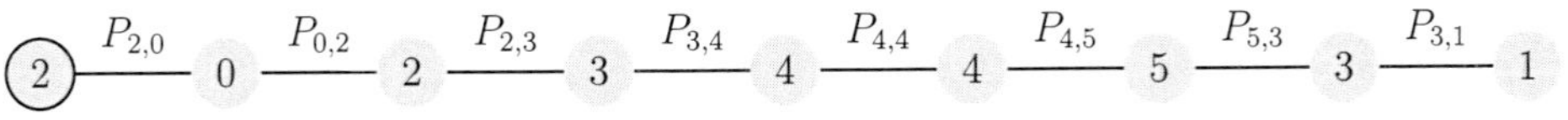

$$\textcircled{2} \overset{P_{2,0}}{\quad\quad} 0 \overset{P_{0,2}}{\quad\quad} 2 \overset{P_{2,3}}{\quad\quad} 3 \overset{P_{3,4}}{\quad\quad} 4 \overset{P_{4,4}}{\quad\quad} 4 \overset{P_{4,5}}{\quad\quad} 5 \overset{P_{5,3}}{\quad\quad} 3 \overset{P_{3,1}}{\quad\quad} 1$$

Figure 5.5: *Example from Figure 5.1. The bold-framed value and the two transition probabilities are needed by the iMc encoder as side information. Optionally, we can apply tree differences, see Section 5.7.1.*

Based on the Tree T we determine the transition frequencies and store them in the CSR matrix format. We choose a sparse matrix format for the transition frequencies, since the number of nonzero entries is less or equal the number of nodes n of tree T. In the case of a huge number of different node values and therefore very big alphabets, we can assure that run time and memory consumption behave linear in the size of the alphabet compared to a quadratic behavior for a dense matrix.

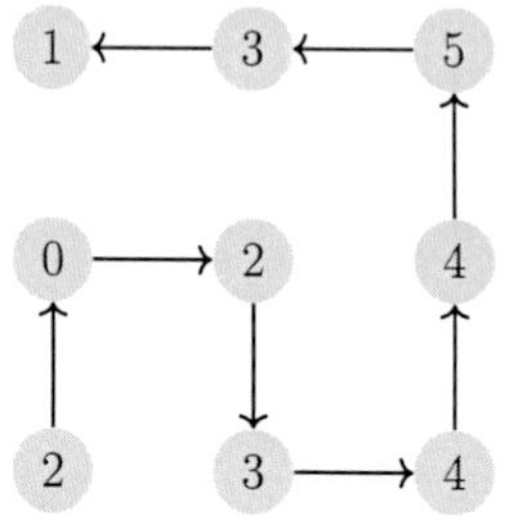

$$P_T^F = \begin{pmatrix} 0 & 0 & 1 & 0 & 0 & 0 \\ 0 & 0 & 0 & 0 & 0 & 0 \\ 1 & 0 & 0 & 1 & 0 & 0 \\ 0 & 1 & 0 & 0 & 1 & 0 \\ 0 & 0 & 0 & 0 & 1 & 1 \\ 0 & 0 & 0 & 1 & 0 & 0 \end{pmatrix}, \quad \begin{aligned} i^{P_T^F} &= [0,1,1,3,5,7,8] \\ j^{P_T^F} &= [2,0,3,1,4,4,5,3] \\ v^{P_T^F} &= [1,1,1,1,1,1,1,1] \end{aligned}$$

Figure 5.6: Determination of the transition frequencies P_T^F for the tree T. For readability the matrix P_T^F is displayed once in a dense matrix form. In our implementation all matrices are described in the CSR sparse matrix format.

For the tree T from Figure 5.6 and its corresponding transition frequency matrix P_T^F, the cumulated frequencies are displayed in Equation (5.4):

$$P_T^F = \begin{pmatrix} 0 & 0 & 1 & 0 & 0 & 0 \\ 0 & 0 & 0 & 0 & 0 & 0 \\ 1 & 0 & 0 & 1 & 0 & 0 \\ 0 & 1 & 0 & 0 & 1 & 0 \\ 0 & 0 & 0 & 0 & 1 & 1 \\ 0 & 0 & 0 & 1 & 0 & 0 \end{pmatrix}, \quad \begin{aligned} i^{P_T^F} &= [0,1,1,3,5,7,8] \\ j^{P_T^F} &= [2,0,3,1,4,4,5,3] \\ v^{P_T^F} &= [1,1,1,1,1,1,1,1] \end{aligned} \tag{5.4}$$

Equation (5.5) shows the formula for cumulated frequencies.

$$(P_T^{F,C})_{ij} = \sum_{k=1}^{j} (P_T^F)_{ij}, \quad \forall i,j = 1,...,\#\mathcal{A}. \tag{5.5}$$

Again, we can exploit the fast access for all non zero elements due to the CSR format.

$$P_T^{F,C} = \begin{pmatrix} 0 & 0 & 1 & 0 & 0 & 0 \\ 0 & 0 & 0 & 0 & 0 & 0 \\ 1 & 0 & 0 & 2 & 0 & 0 \\ 0 & 1 & 0 & 0 & 2 & 0 \\ 0 & 0 & 0 & 0 & 1 & 2 \\ 0 & 0 & 0 & 1 & 0 & 0 \end{pmatrix}, \quad \begin{aligned} i^{P_T^{F,C}} &= [0,1,1,3,5,7,8] \\ j^{P_T^{F,C}} &= [2,0,3,1,4,4,5,3] \\ v^{P_T^{F,C}} &= [1,1,2,1,2,1,2,1] \end{aligned} \tag{5.6}$$

The map matrix is used to accelerate the speed of random access for elements from the cumulated transition frequency matrix $P_T^{F,C}$. For every nonzero element of $P_T^{F,C}$ it has one value that defines on which position of all nonzero elements in one row we can find the number of transitions. For the transition frequency matrices P_T^F and $P_T^{F,C}$ we get the following map matrix:

$$\begin{aligned} i^{P_T^{F,C}} &= [0,1,1,3,5,7,8] \\ j^{P_T^{F,C}} &= [2,0,3,1,4,4,5,3] \\ v^{P_T^{F,C}} &= [1,1,2,1,2,1,2,1] \end{aligned} \qquad \begin{aligned} i^M &= i^{P_T^{F,C}} \\ j^M &= j^{P_T^{F,C}} \\ v^M &= [0,0,1,0,1,0,1,0]. \end{aligned} \tag{5.7}$$

In the equation above, as well as for the CSR matrix all indices and positions are zero-based.

Remark 5.5. *If the alphabet $\mathcal{A}$ contains negative values we perform a shift or an interleaving to bijectively map all values of $\mathcal{A}$ to the set of nonnegative integers $\mathbb{N}_0$.*

Figure 5.7 displays the arithmetic encoding for our Model problem 1. For details of the Arithmetic encoding algorithm we refer to Section 3.3.3.2. For the theoretical investigation of Arithmetic encoding we use transition probabilities instead of transition frequencies to store the stream as a single number in the range $[0, 1)$. By dividing with the row sum, if it is nonzero, we can get the transition probabilities from the transition frequencies.

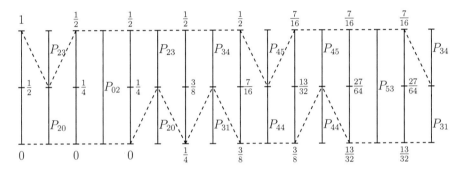

Figure 5.7: Arithmetic coding using transition probabilities.

In our implementation, we apply an Arithmetic encoding using the transition frequencies determined in the fifth step of the iMc encoding, see Figure 5.6.

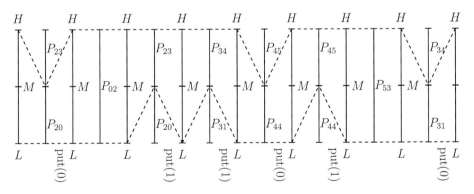

Figure 5.8: Arithmetic coding using the transition frequencies P_T^F. The encoded bit stream is: 011010. The lower and upper bound as well as the center is set to $L = 0$, $H = 2^{31} - 1$, and $M = \left\lceil \frac{H-L}{2} \right\rceil$. The put operation puts the specified bit(s) to the output stream and doubles the size of the interval. This step is called renormalization [93, 115]. Since the frequencies are powers of 2 the interval $[L, H] = [0, 2^{31} - 1]$ stays the same.

Remark 5.6. *The value stream $I = \{2, 0, 2, 3, 4, 4, 5, 3, 1\}$ on the tree of Figure 5.4 can be compressed to 6 Bits "011010" plus the size for the root node value, when the decoder has also access to the applied distribution. The number of compressed bits is distinctly smaller compared to arithmetic encoding based on the relative frequencies, where 23 Bits are used, see Remark 3.68.*

The first two steps of encoding and decoding are identical. Assuming that the encoded values are available the main steps of the decoding algorithm are:

3. decode transition frequency matrix

4. cumulate transition frequencies, see (5.5) (5.4)

5. arithmetic decoding, see Figure 5.9

6. tree and time sums (if necessary), see Section 5.7.1.

Figure 5.9 shows the iMc decoding process. We apply the transition frequencies and the root node value as side information to decode the bit stream "011010" from Figure 5.8. The decoding process is similar to the encoding process. The only difference is that we determine the interval based on the bit stream instead of the values to encode.

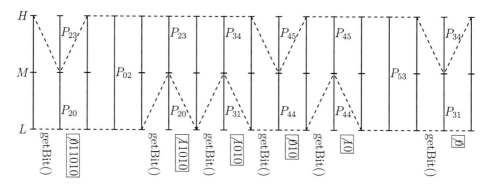

Figure 5.9: *Arithmetic decoding using the stored transition frequencies P_T^F. The encoded bit stream is: 011010. The current bit buffer is available in the rectangular boxes. The lower and upper bound as well as the center is set to $L = 0$, $H = 2^{31} - 1$, and $M = \left\lceil \frac{H-L}{2} \right\rceil$. The getBit operation takes as many bits as it needs to specify the interval for the next letter to decode. Since the cumulated frequencies are 1 or 2, we need only one bit per letter. With the cumulated frequencies being powers of 2 interval $[L, H] = [0, 2^{31} - 1]$ remains the same.*

We compress the iMc statistics lossless by exploiting that in an arithmetic encoder we use integers, i.e. the frequencies, to describe the probabilities. For the structure of the transition probability matrix we use the *Compressed Sparse Row* (CSR) format. All integer arrays are decorrelated and encoded with zlib [111] afterwards.

The construction of the spanning tree based on the adjacency matrix ensures that we get the identical tree even if the elements are in a different order at decompression.

Remark 5.7. *There was no implementation of an arithmetic encoder available that fulfills all requirements for an application in our iMc encoder. One-pass encoders, such as DMC or PPM, which specify the statistics on the fly, cannot omit zero probabilities, see Section 3.3.3. The property to omit zero probabilities is crucial since, in general, we handle quantized data as single precision integer values. The size of the complete alphabet is 2^{32} while only a small portion has nonzero probability. An application of PPM and DMC on integer data with the need of specifying nonzero values for all values from the alphabet results in bad compression rates. In general, PPM and DMC is applied bitwise or byte-wise to avoid nonzero probabilities. Moreover building up transition probabilities on the fly is computationally expensive. We, therefore, developed an arithmetic encoder following the basic implementation of [11] using the techniques of [115].*

An important property of the iMc encoder is that it omits zero probabilities. This allows us to create the statistics for 4 bytes integer data in contrast to bits and bytes for PPM and DMC. Since the alphabet size for 4 bytes integer data is 2^{32}, no meaningful compression is possible without omitting zero probabilities. For the construction of the transition probabilities, we need to use 4 bytes integer data as we want to apply the iMc encoder as a black box tool and we expect cases where the data types like 2 bytes integer data is not sufficient.

Our application field is predestinated for an application of asymmetric compression, see Section 2.2. We, therefore, apply a two pass encoder. In the first pass we gather the information needed to determine the statistics. In the second pass we encode the stream using the precalculated statistics. For decoding, we only have one pass as the statistics are stored in the compressed stream.

5.2   Time complexity of iMc encoding

In this section, we estimate the time complexity of the iMc encoder in our application field first for general graphs and second for graphs based on finite element grids. Next we compare the time complexity of iMc to those of Markov Random Fields. We split our investigation on the time complexity of our iMc encoder into the following four stages:

- initialization

- prediction and inverse-prediction

- encoding

- decoding

In the initialization stage, we determine a tree based on the finite element grid of the investigated part. This step has to be performed at the time of encoding, if the tree is not yet available. For the decoding stage, it is only necessary, if the tree is not stored as part of the compressed data stream. The prediction is optional and not mandatory in our iMc encoder. Nevertheless, in general it is meaningful to apply a prediction such as tree differences, see Section 3.3.2.3. If we use a tree for prediction, we can use the same tree for determining the statistics in order to save run time.

Table 5.1: Time complexity of the iMc encoder and decoder.

Initialization	
1. Adjacency matrix of graph	$\mathcal{O}(e \cdot (log_2(deg) + 1) + n)$
2. Find tree (Trémaux)	$\mathcal{O}(e)$
Compression	
3. Prediction (optional)	$\mathcal{O}(n \cdot ts)$
4. Adjacency matrix of tree	$\mathcal{O}(n \cdot (\log_2(deg) + 1))$
5. Transition frequencies	$\mathcal{O}(n \cdot ts \cdot (\log_2(maxP) + 1))$
6. Cumulate frequencies	$\mathcal{O}(nnzP)$
7. Construct map matrix	$\mathcal{O}(nnzP)$
8. Arithmetic encoding	$\mathcal{O}(n \cdot ts \cdot (\log_2(maxP) + 1))$
9. Encoding side information	$\mathcal{O}(nnzP + ts \cdot root)$
Decompression	
3. Decode side information	$\mathcal{O}(nnzP + ts \cdot root)$
4. Cumulate frequencies	$\mathcal{O}(nnzP)$
5. Arithmetic decoding	$\mathcal{O}(n \cdot ts \cdot (\log_2(maxP) + 1))$
6. Reverse prediction (optional)	$\mathcal{O}(n \cdot ts)$

Variable	Description
deg	maximal degree of all nodes in G
n	number of nodes in G, $n = \#V$
e	number of edges in G, $e = \#E$
$nnzP$	number of nonzero elements of the transition probability matrix P
ts	number of points in time
$maxP$	maximal number of nonzero entries per line in matrix P
m	size of the alphabet $\mathcal{A}$
$root$	number of root nodes in tree

Theorem 5.8. *Let $G = (V, E)$ be a connected graph and $\mathcal{A} = \{a_1, ..., a_m\}$ an alphabet. Furthermore, let the maximal degree of a node in G be denoted by deg, the number of nodes by n, the number of edges by e, the number of nonzero elements of the transition probability matrix by $nnzP$, the number of points in time stored in the simulation result file by ts, the maximum number of entries per line in the transition probability matrix by $maxP$, and the number of root nodes by $root$. The time complexity for the initialization stage is*

$$T_{init} = \mathcal{O}(e \cdot (\log_2(deg) + 1) + n) \tag{5.8}$$

and for the tree prediction T_{pred} and its inverse prediction $T_{invpred}$, it is

$$T_{pred} = T_{invpred} = \mathcal{O}(n \cdot ts) \tag{5.9}$$

The time complexity for the encoding stage is

$$T_{enc} = \mathcal{O}(n \cdot (\log_2(deg)+1)) + 2\mathcal{O}(n \cdot ts \cdot (\log_2(maxP)+1)) + 3\mathcal{O}(nnzP) + \mathcal{O}(root \cdot ts) \tag{5.10}$$

and for the decoding stage, it is

$$T_{dec} = \mathcal{O}(n \cdot ts \cdot (\log_2(maxP) + 1)) + \mathcal{O}(nnzP) + \mathcal{O}(root \cdot ts). \tag{5.11}$$

Table 5.1 shows an overview of all variables $deg, n, e, nnzP, ts, maxP, root$.

PROOF: We start with the time complexity of initialization stage. In this stage we create the tree from the finite element description of the mesh.

1. Adjacency matrix of graph:

Since adding nonzero elements to a matrix in CRS format is computationally expensive and we are interested on the position of nonzero elements, we apply a sorted list for every row of the matrix. Building up this structure needs $\mathcal{O}(n)$ operations. Inserting one edge in the sorted list uses not more than $\log_2(deg) + 1$ operations, i.e. $\mathcal{O}(e(\log_2(deg) + 1))$. After processing all edges, we transform the adjacency matrix into the CRS format, which consumes one operation per nonzero element of the adjacency matrix $nnzA$. The number $nnzA$ is twice the number of edges if there are no self-connected nodes and less otherwise. Therefore, transforming the adjacency matrix into the CRS format is $\mathcal{O}(e)$.

2. Find tree:

The time complexity for finding a depth first search tree is depended on the used algorithm. For Trémaux's algorithm, the run time is $\mathcal{O}(e)$ [42] in the worst case scenario. Being undirected and no loops are allowed, the tree has not more than $n - 1$ edges. The equilibrium holds if the graph is connected.

The overall time complexity of the initialization phase is:

$$T_{\text{init}} = \mathcal{O}(e \cdot \log_2(deg) + n) + \mathcal{O}(e)$$
$$= \mathcal{O}(e \cdot (\log_2(deg) + 1) + n)$$

The application of tree prediction is optional and investigated in the following.

3. Prediction:

When we apply tree prediction, we have to process each non-root element once for each time step. Therefore, the time complexity of prediction step is not more than $\mathcal{O}(n \cdot ts)$, see Section 3.3.2.3. While processing, we also determine the minimum and maximum of the predicted values.

For compression we have, next to the initialization and the optional prediction, the following time complexity for the encoding step.

4. Adjacency matrix of tree:

Since the tree has a maximum of $n - 1$ edges and each node has not more than deg neighbors, building up its adjacency matrix does not take more than $\mathcal{O}(n \cdot (\log_2(deg) + 1))$ operations.

5. Transition probabilities:

To build up the transition probability matrix we have to process each non-root node value once for each time step: $\mathcal{O}(n \cdot ts)$. Same as for the adjacency matrix we use sorted lists to build up the nonzero structure. In this case the value vector of the transition probabilities in CSR matrix format is required. We have to apply the same sorting for the value vector as for the column index list. For each edge of the tree the corresponding node value pair $(a_k, a_l) \in \mathcal{A} \times \mathcal{A}$ is inserted into the transition probability matrix Each insertion takes $\mathcal{O}(\log_2 maxP + 1)$. Since there are not more than $n - 1$ edges in each time step, we have an overall run time of $\mathcal{O}(n \cdot ts \cdot (\log_2(maxP) + 1))$

6. Cumulate frequencies:

Cumulating the frequencies is required for the implementation of our arithmetic encoder.

It sums up the number of transitions per line of matrix P and stores it in CRS format, i.e. $\mathcal{O}(nnzP)$ operations.

7. Construct map matrix:

The map matrix has the same structure as the transition probability matrix. They only differ for the value vector, which represents the position of a nonzero value in the column index vector. Building up this vector consumes $\mathcal{O}(nnzP)$ operations.

8. Arithmetic encoding:

For each element that will be encoded, i.e. every non-root element for every time step, we need $\mathcal{O}(maxP)$ operations to get the lower and upper bound of the interval for encoding, see Section 3.3.3.2. The encoding process itself needs some bit operations, whereas the overall number can be approximated by a small constant.

9. Encoding side information:

In the last step, we encode the statistics as well as all root node values for all time steps, i.e. $\mathcal{O}(nnzP + ts \cdot root)$

We can sum up the time complexities for all steps of the encoding stage and get the time complexity of Equation (5.10):

$$
\begin{aligned}
T_{enc} =& \mathcal{O}(n \cdot (\log_2(deg) + 1)) + 2\mathcal{O}(n \cdot ts \cdot (\log_2(maxP) + 1)) + 2\mathcal{O}(nnzP) \\
& + \mathcal{O}(nnzP + ts \cdot root) \\
=& \mathcal{O}(n \cdot (\log_2(deg) + 1)) + 2\mathcal{O}(n \cdot ts \cdot (\log_2(maxP) + 1)) + 3\mathcal{O}(nnzP) + \mathcal{O}(ts \cdot root)
\end{aligned}
$$

For decompression we have the following time complexity. The additional time complexity for the initialization phase is only necessary in cases, where the tree is not available.

3. Decoding side information:

The time complexity is the same like for encoding the statistics.

4. Cumulate frequencies:

The time complexity is the same like for cumulating the frequencies in compression phase.

5. Arithmetic decoding:

The time complexity is the same like for arithmetic encoding in compression phase, see Section 3.3.3.2.

When we sum up the time complexities for all steps of the decoding stage, we get the time complexity of Equation (5.11):

$$
\begin{aligned}
T_{dec} =& \mathcal{O}(n \cdot ts \cdot (\log_2(maxP) + 1)) + \mathcal{O}(nnzP) + \mathcal{O}(nnzP + ts \cdot root) \\
=& \mathcal{O}(n \cdot ts \cdot (\log_2(maxP) + 1)) + 2\mathcal{O}(nnzP) + \mathcal{O}(ts \cdot root)
\end{aligned}
$$

If a tree prediction was applied on the data stream, an inverse prediction has to be applied, whose time complexity is investigated in the following.

6. Reverse prediction:

The time complexity is the same like for prediction in compression phase, see Section 3.3.2.3. □

In general, the difference in run time for the encoding as well as the decoding stage is dominated by the term $\mathcal{O}(n \cdot ts \cdot (\log_2(maxP) + 1))$. We, therefore, expect that the data can be decompressed with twice the speed of compression.

Remark 5.9. *When we compare, T_{enc} and T_{dec}, we see that the time complexity of the encoding is distinctly higher than the one of decoding. Therefore, the iMc encoder is* *an asymmetric encoder, see Definition 3.50. Moreover, since we traverse the data twice while encoding for building up the transition frequencies and for the arithmetic encoding, we have a two-pass encoder.*

For a part with shell elements the average number of neighbors is approximately four. The maximal degree of a node deg for 2D and 3D elements in the crash context is usually bounded by 6. Thus, $\log_2(deg)$ can be neglected. A shell element part from a crash simulation is usually connected and if it is not, consists of only few connected components. Therefore, the number of root nodes can be neglected as well. For car crash simulations between 25 and 60 time steps are kept for further postprocessing. Therefore, all operations on the finite element grid that are performed only once can be neglected, e.g. the initialization, the creation of the adjacency matrix of the tree, the cumulation of frequencies and the construction of the map matrix. Since the number of root node values $root$ is also small, the encoding and decoding of side information can be neglected as well. For compressing a set of simulation results the tree has to be only calculated once for all simulation results processed at the same time. When we apply the iMc encoder on the residual of PPCA prediction, see Section 4.3, the number of time steps coincides with the sum of all time steps of all processed simulation results. Furthermore, in the case of PPCA we can store the tree as part of the database and the initialization has to be done only once during the compression. The tree can be reused for all PPCA online steps, see Section 4.3.3, as well as for decompression. Hence, the initialization can be neglected in the case of PPCA.

With shell elements of car crash simulations the time complexity of encoding as well as decoding is dominated by the term $\mathcal{O}(n \cdot ts \cdot (\log_2(maxP)+1))$. Since we encode a symbol for all nodes and all time steps lossless, the run time is at least $\mathcal{O}(n \cdot ts)$. The crucial parameter is $\log_2(maxP)$, which is the logarithm to base 2 of the maximal number of nonzero elements in transition frequency matrix P. The parameter $maxP$ gives us the maximal number of possible transitions from one symbol of the alphabet. Therefore, it holds that $maxP \leq m$. If the alphabet is small, e.g. after prediction, the time complexity is close to its lower bound. In the case of a big alphabet, $maxP$ can be small, too, e.g. in the case of a small maximal degree deg, and the connected nodes have similar node values. In contrast, $maxP$ is big if deg is very high, e.g. close to n. This is the case if one node is connected with all other nodes. The dependency of the iMc encoder time complexity on $maxP$ is one reason why we use a depth first search algorithm instead of breadth first search, since the possibility that we have less transitions from value increases, if the number of different start nodes of edges increases.

We, finally, compare the time complexity of the iMc encoder with the one of Markov Random Fields (MRF), see Paragraph 3.3.3.7.

Considering only one state of the simulation, the time complexity can be bounded by $\mathcal{O}(n \cdot (\log_2(maxP)+1))$. By construction the maximal number of nonzero entries per row of P $maxP$ is not bigger than the size of the alphabet $\mathcal{A}$, i.e. $maxP < m$. Therefore, iMc encoding is applicable in the case of a large alphabet, especially when the number of possible transitions from one value to another is small. When the data is predicted successfully the size of the alphabet is small and can be bound by 500. Therefore, for big meshes the run time is linear in the number of edges.

When we consider an entropy encoding scheme with MRFs on a spanning tree for one variable and one point in time the time complexity of generating the statistics is $\mathcal{O}(n \cdot m^2)$, see Remark 3.77 and [108]. Regarding time complexity, opposite to MRF-encoding schemes the iMc can process large alphabets which allows it to be implemented as a black box tool.

5.3 Adaptive Arithmetic coding using iMc

In this section, we introduce the Adaptive iMc encoder that exploits the knowledge regarding all previously processed elements. An adaptive compression method has the property that it can change the strategy how to compress a data set based on the properties of the data set, see Definition 3.49.

The iMc encoder introduced in Section 5.1 exploits the knowledge based on induced Markov chains reflecting the underlying distribution. Since we store the number of transitions and have a serial execution of encoding and decoding, we can exploit information on events already processed. The idea is to eliminate processed events from the distribution and use the updated statistics in both the encoding and the decoding stage. Because we exploit more information in the encoding step about previously encoded transitions, the compression rates for the adaptive iMc encoding are at least as good as for the non-adaptive approach of Section 5.1. The optimized compression rate is achieved by investing additional computations when updating the statistics. We, again, have two ways to describe our approach. The first one uses transition probabilities and can be seen as a theoretical approach, cf. Table 5.2. The second one is based on the modification of the transition frequencies being part of our implementation.

Lemma 5.10. *Let an alphabet* $\mathcal{A} = \{a_1, ..., a_m\}$ *and a tree* $T = (V_T, E_T)$ *be given and the current letter be* a_i, $i \in \{1, ..., m\}$. *Moreover, let the next observed transition be from* a_i *to* a_j *with* $j \in \{1, ..., m\}$. *The effect of eliminating the event to go from* a_i *to* a_j *modifies the transition frequency matrix as follows,*

$$(P_T^F)_{ij} = (P_T^F)_{ij} - 1.$$

The cumulated frequency matrix has to be updated like

$$(P_T^{F,C})_{ik} = (P_T^{F,C})_{ik} - 1, \quad \forall k = j, ..., m.$$

After processing all edges, for all i, j *it holds*

$$(P_T^F)_{ij} = (P_T^{F,C})_{ij} = 0$$

For the adaptive iMc encoder, it is necessary to perform updating in both encoding and decoding.

PROOF: For $i, j \in \{1, ..., m\}$ the transition frequencies are defined as $(P_T^F)_{ij} = \#\{(k, l) \in E_T | v_k = a_i \wedge v_l = a_j\}$, see Definition 5.2. Processing the edge $(k, l) \in E_T$ with node values $v_k = a_i$ and $v_l = a_j$ we observe the event (a_i, a_j). Generating the statistics for compression we set $(P_T^F)_{ij} = (P_T^F)_{ij} + 1$ and continue traversing the tree. Finally, this results in a transition frequency matrix $(P_T^F)_{ij}$ for all $i, j = 1, ..., m$ that consists of the number of transitions made from $a_i \in \mathcal{A}$ to $a_j \in \mathcal{A}$. At the time of decompression we set

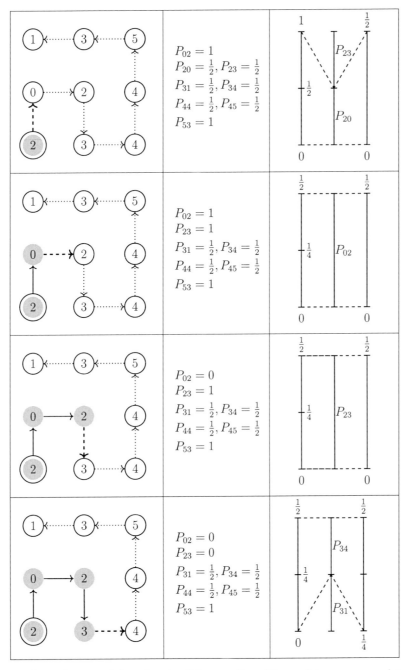

Table 5.2: Several steps of the adaptive iMc encoding. The update process and its effect on the encoding are shown using transition probabilities.

$(P_T^F)_{ij} = (P_T^F)_{ij} - 1$ after processing a transition from $a_i \in \mathcal{A}$ to $a_j \in \mathcal{A}$ and continue the transition. With the compression being lossless and all events that take place while compression also occurring at the same place during decompression, we get $(P_T^F)_{ij} = 0$ for all i, j.

The cumulated frequencies are determined by $(P_T^{F,C})_{ij} = \sum_{k=1}^{j}(P_T^F)_{ij}$ for all $i, j = 1, ..., m$, see Equation (5.5). When we update the frequency $(P_T^F)_{ij} = (P_T^F)_{ij} - 1$, we also have to update all entries of the cumulated frequency matrix $(P_T^{F,C})_{ik}$ with $k = j, ..., m$. We already saw that all transition frequencies are zero after processing all events. Therefore the sum of these frequencies will also be zero and therefore $(P_T^{F,C})_{ij} = 0$ holds.   □

If the number of all possible transitions are high, updating the cumulated frequencies can be computationally expensive. A solution is to collect all events and update the statistics not every time but all 100th time.

Remark 5.11. *The adaptive iMc encoder is distinguished from other adaptive encoding strategies such as PPM and DMC, see Sections 3.3.3.3 and 3.3.3.5 by the following criterion: Whereas other approaches adapt their statistics based on already processed elements the adaptive iMc encoder adapts its statistics with respect to the values that will be encoded and decoded in the future. Changing the statistic based on future data is possible since the iMc is aware of all transitions and save this as side information for decoding.*

Adaptive iMc encoding gains better compression rates than the standard iMc encoding, when the number of transitions leaving a letter is small compared to the number of all transitions. In such a case, subtracting a single event from the statistic can change the probabilities drastically. If after one update step several thousand events remain, the overall effect on updating the statistic is rather small. Therefore, the effect on the compression rate is also small. The adaptive iMc is promising in the following scenarios:

- Small graphs,

- Badly and non predicted data sets.

The overall adaptive encoding of the Model problem 1 using the spanning tree of Figure 5.3 is displayed in Figure 5.10. Therefore, we only need the three bits "010" next to the side

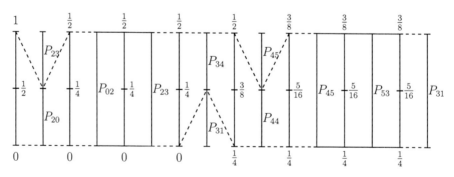

Figure 5.10: *Adaptive iMc coding of Model Problem 1 with the spanning tree of Figure 5.3.*

information of root node value and statistics to encode the example of MP 1. Compared to the non-adaptive encoding displayed in Figure 5.7, we only need three instead of six bits. Unfortunately, in our application area, we expect big graphs with several hundred up to many thousand nodes and well predicted data sets.

The distribution of the adaptive iMc encoder is dependent on its position on the graph, i.e. on the already processed node values. Therefore, the underlying Markov model is inhomogeneous, see Remark 3.72.

5.4 Theory of induced Markov chains

In this section, we introduce induced Markov chains that are the mathematical background of the iMc encoder, see Section 5.1. We, moreover, specify the set of graphs that induces mean ergodic and ergodic Markov chains.

Mean ergodicity is necessary for a well-defined entropy that represents the best possible compression rate, see Corollary 3.43 and Theorem 3.20. Ergodicity generally means the property that the overall behavior is independent from the first events, whereas for mean ergodicity the overall averaged behavior is independent. Since the convergence speed of the Markov chain to its stationary distribution is higher in the case of ergodicity compared to the mean ergodic case, see Remark 3.46 and we determine the entropy via the stationary distribution, we try to proof entropy where ever possible. To improve the readability basic definitions in the context of probability theory are moved to Appendix A.1.1.

We start with a random walk on a graph.

Definition 5.12 (random walk [86]). *A random walk is a Markov chain with a finite state space that is time-reversible.*

A grid of a finite element simulation can be interpreted as an undirected graph $G_u = (N, E_u)$. The set of node values forms alphabet $\mathcal{A}$. We are interested in the probability having a node with value $i \in \mathcal{A}$ together with a neighbored node with value $j \in \mathcal{A}$. We gather all probabilities as a set of distributions and can use this as side information in a statistical encoder, see Section 5.1.

Since we are interested in how probable it is to be at a node with a certain value and go to a neighboring node with a possibly different value, it is necessary to know the direction of the edge. Having an undirected connected graph $G_u = (N, E_u)$ we determine a directed graph $G = (N, E)$ by duplicating each edge of E_u and orient them both ways, see Figure 5.11. Let the number of nodes be $n = \#N$ and the number of directed edges be $e = \#E$. Furthermore, let a finite alphabet $\mathcal{A} = \{a_1, ..., a_m\}$, and a node value vector $v = (v_1, ..., v_n) \in \mathcal{A}^n$ be given.

Remark 5.13. *Without loss of generality we assume that all elements of $\mathcal{A}$ are part of the node value vector v. In this case we call $\mathcal{A}$ minimal.*

The minimality is not necessary for the algorithm, but simplifies the later.

Let $(N, \mathcal{P}(N), \mathbb{P}_N)$ be the *Laplace probability space of order n*, see Definition A.7, where all n elementary events are equiprobable, i.e. $\mathbb{P}_N[n_i] = \frac{1}{n}$. Since $\mathcal{A}$ is finite, the tuple $(\mathcal{A}, \mathcal{P}(\mathcal{A}))$ is a measurable space. We define a random variable $Y : N \to \mathcal{A}$ with $Y : n_i \mapsto a_l$ that maps a node to the value of the alphabet, which is defined by an entry of the node value vector v. We identify v as a result of the random process.

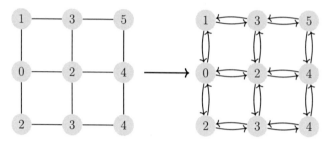

Figure 5.11: Interpretation of the graph G from Model Problem 1 as a directed graph.

Definition 5.14 (probability mass function [128]). *Let Y be a random variable with finite state space* $\mathcal{A}$. *The probability mass function for Y provides the probability that the value obtained by Y on the outcome of an experiment is equal to* $a \in \mathcal{A}$.

For a random variable Y with probability space $(N, \mathcal{P}(N), \mathbb{P}_N)$ and finite state space $\mathcal{A}$ the probability mass function is defined by $\mathbb{P}_Y := \mathbb{P}_N \circ Y^{-1}$ [17].

$$\mathbb{P}_Y[a_l] = \mathbb{P}_N[\{n_i \in N | Y(n_i) = a_l\}]$$
$$= \frac{1}{n} \#\{n_i \in N | Y(n_i) = a_l\}. \tag{5.12}$$

We obtain the empirical probabilities, see Definition 5.1, as the probability mass function of the random variable Y. The distribution coincides with the relative frequencies of the node values. In the following we use the term initial distribution instead of empirical probabilities to underline that it is a distribution:

Definition 5.15 (initial distribution). *The initial distribution* ν *is defined by the relative frequency of the node values:*

$$\nu_i := \mathbb{P}_Y[a_i] = \frac{\#\{k \in N | v_k = a_i\}}{n}. \tag{5.13}$$

The following theorem describes the theoretical background of the iMc encoder.

Theorem 5.16. *Let* $(E, \mathcal{P}(E), \mathbb{P}_E)$ *be a Laplace probability space of order* e, $X : E \to \mathcal{A} \times \mathcal{A}$ *with* $X : e_i \mapsto (a_k, a_l)$ *be a two dimensional random variable, and* Z *be a random variable that obtains the node values when performing a random walk on a graph* G. *Then the conditional distribution* $\mathbb{P}_{X|Z}$ *coincides with the transition probabilities from Equation (5.2):*

$$\mathbb{P}_{X|Z}[(a_k, a_l)|a_k] = P_{kl}, \quad \forall k, l \in \{1, ..., m\}. \tag{5.14}$$

PROOF: We define a two dimensional random variable $X : E \to \mathcal{A} \times \mathcal{A}$ with $X : e_i \mapsto (a_k, a_l)$. The probability mass function for the random variable is defined by $\mathbb{P}_X = \mathbb{P}_E \circ X^{-1}$:

$$\mathbb{P}_X[(a_k, a_l)] = \mathbb{P}_E[\{e_i \in E : X(e_i) = (a_k, a_l)\}]$$
$$= \frac{1}{e} \#\{e_i \in E : X(e_i) = (a_k, a_l)\} \tag{5.15}$$

$$= \frac{1}{e}\#\{(i,j) \in E | v_i = a_k \wedge v_j = a_l\}. \tag{5.16}$$

With the help of the two distributions $\mathbb{P}_Y$ and $\mathbb{P}_X$ we can imagine the statistical model as follows. We start with a graph with no node values and randomly choose one node with a uniform distribution. We then assign a value from the alphabet to this node depending on the relative frequency of the node values, cp. Equation (5.12). Next, we start a random walk at the selected node with the chosen node value. For a random walk on a connected graph, every edge is equiprobable if the first edge was chosen equiprobable. If not, the probability mass function of the edges converges with the number of steps of the random walk to a uniform distribution [86]. We, therefore, also use a Laplace probability space for selecting the edges, see Equation (5.15). We then assign a value to each node depending on the transition probabilities of the given graph G as specified next.

Based on the graph's structure we can determine how often a node will be reached in a random walk on the graph and thus how often a certain value of the alphabet will be reached. For a random walk on a graph with equiprobable edges the probability to be at node n_i is

$$\mathbb{P}[n_i] = \frac{deg(n_i)}{e}, \tag{5.17}$$

whereby $deg(n_i)$ represents the *degree* of node n_i, i.e. the number of neighbors [86]. Let Z be a random variable that takes the node values for a random walk on a graph G. Hence, considering Equation (5.17), the probability for a certain value to be the first entry of a directed edge value tuple is

$$\mathbb{P}_Z[a_k] := \frac{1}{e} \sum_{\{n_i \in N | Y(n_i) = a_k\}} deg(n_i)$$

$$= \sum_{a_l \in \mathcal{A}} \mathbb{P}_X[a_k, a_l] \tag{5.18}$$

$$= \frac{1}{e}\#\{(i, \cdot) \in E | v_i = a_k\}.$$

If the measure $\mathbb{P}_X$ is seen as a joint distribution of twice the random variable Y with probability mass function $\mathbb{P}_Y$, the equation (5.18) also represents the marginal distribution for a certain starting node and all possible end node values in $\mathcal{A}$. Based on the joint distribution (5.15) and the marginal distribution (5.18), we can determine the conditional probabilities:

$$\mathbb{P}_{X|Z}[(a_k, a_l)|a_k] = \frac{\mathbb{P}_X[(a_k, a_l)]}{\mathbb{P}_Z[a_k]}$$

$$= \frac{\#\{(i,j) \in E | v_i = a_k \wedge v_j = a_l\}}{\#\{(i, \cdot) \in E | v_i = a_k\}}. \tag{5.19}$$

$\square$

Theorem 5.17. *The marginal distribution $\mathbb{P}_Z$ from Equation (5.18) of a finite element alphabet $\mathcal{A}$ is a stationary distribution in regard to the transition probability matrix P from Equation (5.14).*

PROOF: Let P be a stochastic matrix. A distribution π is stationary in regard to P if and only if $\pi P = \pi$, see Definition 3.27. Let an alphabet $\mathcal{A} = \{a_1, ..., a_m\}$, the marginal distributions $\pi = (\mathbb{P}_Z[a_k])_{k=1,\ldots,m}^T$, and the conditional distribution $P = (\mathbb{P}_{X|Z}[(a_k, a_l)|a_k])_{k,l=1,\ldots,m}$ be given. Then

$$\pi P = \left((\mathbb{P}_Z[a_k])_{k=1,\ldots,m}^T (\mathbb{P}_{X|Z}[(a_k, a_l)|a_k])_{k,l=1,\ldots,m} \right)$$

$$= \left(\left(\frac{1}{e} \#\{(i,\cdot) \in E | v_i = a_k\} \right)_{k=1,\ldots,m}^T \left(\frac{\#\{(i,j) \in E | v_i = a_k \wedge v_j = a_l\}}{\#\{(i,\cdot) \in E | v_i = a_k\}} \right)_{k=1,\ldots,m} \right)_{l=1,\ldots,m}$$

$$= \frac{1}{e} \left(\sum_{k=1}^{m} \#\{(i,\cdot) \in E | v_i = a_k\} \frac{\#\{(i,j) \in E | v_i = a_k \wedge v_j = a_l\}}{\#\{(i,\cdot) \in E | v_i = a_k\}} \right)_{l=1,\ldots,m}$$

$$= \frac{1}{e} \left(\sum_{k=1}^{m} \#\{(i,j) \in E | v_i = a_k \wedge v_j = a_l\} \right)_{l=1,\ldots,m}$$

$$= \frac{1}{e} \left(\#\{(\cdot,j) \in E | v_j = a_l\} \right)_{l=1,\ldots,m}$$

$$= \frac{1}{e} \left(\#\{(j,\cdot) \in E | v_j = a_l\} \right)_{l=1,\ldots,m}$$

$$= \pi.$$

The last to last equation holds since the number of edges pointing to a node with a certain value is equal to the number of edges leaving this node.   □

For an example, see Figure 5.12. We organize the probabilities P_{kl} in the transition

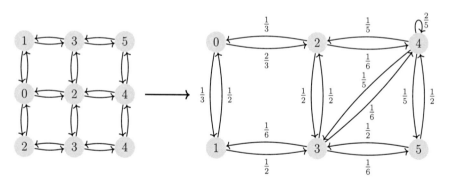

Figure 5.12: Mesh with node values and the induced Markov chain with transition probabilities.

probability matrix P.

Lemma 5.18 ([130]). *The tuple (P, ν) of a stochastic matrix P and a distribution ν define a finite state space and time-homogeneous Markov chain M with transition probabilities given in P and an initial distribution ν.*

$$\mathbb{P}[X_0 = a_i] = \nu_i, \ \mathbb{P}[X_{k+1} = a_j | X_k = a_i] = P_{ij}, \ \forall a_i, a_j \in \mathcal{A}.$$

We introduce the *induced Markov chain* (iMc).

Definition 5.19 (induced Markov chain). *We call the Markov chain from Lemma 5.18 with the transition probabilities defined by Formula (5.14) and the initial distribution defined by (5.13) induced Markov chain.*

Next we investigate the properties of such induced Markov chains. The graphs considered in this thesis have a finite number of nodes and thus a finite alphabet $\mathcal{A}$. A graph based on the finite element mesh is given for one dataset and no predetermined strategy to walk through the graph exists. From the topology point of view we, therefore, assume time-homogeneity for the induced Markov chain, see Definition 3.23. Hence all Markov chains induced by a graph are time-homogeneous, finite-state and have an underlying finite-alphabet Markov source. In the case of adaptive iMc coding, see Section 5.3, we do not have time-homogeneity. The iMc can be seen as a special case of Bayesian networks, see Section 3.3.3.7.

In Section 3.1, we argue that mean ergodicity is crucial for a well-defined entropy determination. Theorem 5.20 proves mean ergodicity for a big pile of data sets only based on the topology of the graph.

In the following, we define classes of graphs that can be identified with ergodic or mean ergodic Markov chains. From Corollary 3.41, we know that in the case of a finite state space, it is sufficient to show irreducibility to get mean ergodicity. If we can also show aperiodicity, the induced Markov chain will even be ergodic.

Theorem 5.20. *A Markov chain M which is induced by a connected undirected graph G is irreducible.*

For the proof of Theorem 5.20 we need a graph-theoretical definition of a walk.

Definition 5.21 (walk [32]). *A walk is a sequence of edges $e_1, ..., e_k \in E$ from an edge set E where the ending node of the predecessor edge is the starting node of the successor edge. Let $(n_{i_{j-1}}, n_{i_j}) = e_j$ for $j = 1, ..., k$. The walk will be denoted by the series of nodes that is traversed $\langle n_{i_0}, n_{i_1}, n_{i_2}, ..., n_{i_{k-1}}, n_{i_k} \rangle$.*

Lemma 5.22. *Let $G = (V, E)$ be an undirected connected graph with $V, E \neq \emptyset$ and $n_i \in V$ a node with node value $s \in \mathcal{A}$ and $n_j \in V$ with $(n_i, n_j) \in E$. There exists a walk $\langle n_i, n_j, n_i \rangle$ and for all $l \in \{2k | k \in \mathbb{N}\}$ it holds*

$$\mathbb{P}[X_l = s | X_0 = s] > 0. \tag{5.20}$$

PROOF: Let $G = (V, E)$ be an undirected connected graph with $V, E \neq \emptyset$ and $n_i \in V$ and $n_j \in V$ be neighbored nodes, i.e. $(n_i, n_j) \in E$ with node values $s, t \in \mathcal{A}$. Then there exists a walk $\langle n_i, n_j, n_i \rangle$ from a node to its neighbor and back to the starting node. Due to the definition of the transition probabilities if $\mathbb{P}[X_1 = t | X_0 = s] > 0$ it follows that $\mathbb{P}[X_1 = t | X_0 = s] > 0$. Thus

$$\mathbb{P}[X_2 = s | X_0 = s] \geq \mathbb{P}[X_1 = t | X_0 = s] \mathbb{P}[X_2 = s | X_1 = t] > 0 \tag{5.21}$$

The probability $\mathbb{P}[X_2 = s | X_0 = s]$ is greater than $\mathbb{P}[X_1 = t | X_0 = s] \mathbb{P}[X_2 = s | X_1 = t]$ if there exists more than one three-node-walk with $X_0 = X_2 = s$. □

It follows the proof of Theorem 5.20.

PROOF: Let $v_1, ..., v_n \in V$ be the node values of the graph G and $\mathcal{A} = \{v_i | i \in \{1, ..., n\}\}$

the set of possible node values. Using the definition of irreducibility, see Definition 3.30, it is sufficient to show: $\forall s, t \in \mathcal{A} : \exists i \in \mathbb{N} : \mathbb{P}[X_i = t | X_0 = s] > 0$.
As the graph G is connected, for all node tuple $(n_i, n_j) \in V \times V$ there exists at least one walk $\langle n_i = n_{l_0}, n_{l_1}, ..., n_{l_{k-1}}, n_{l_k} = n_j \rangle$ with $\langle n_{l_i}, n_{l_{i+1}} \rangle \in E$ for all $l = 0, ..., k-1$. We distinguish three cases. In the first case, the graph consists of only one node with node value $a_1 \in \mathcal{A} = \{a_1\}$. The transition probabilities are $\mathbb{P}_{X|Z}[X_1 = a_1 | X_0 = a_1] = 1$ and the initial distribution is $\mathbb{P}_Y[X_0 = a_1] = 1$. Therefore, $\mathbb{P}_{X|Z}[X_i = a_1 | X_0 = a_1] = 1$ for all $i \in \mathbb{N}$ and the Markov chain is irreducible. In the second case, let $s = v_i$ and $t = v_j$ with $i \neq j$. Then there exists a walk $\langle n_i = n_{l_0}, n_{l_1}, ..., n_{l_{k-1}}, n_{l_k} = n_j \rangle$ with the corresponding node values $v_i, v_{l_1}, ..., v_{l_{k-1}}, v_j \in \mathcal{A}$.
As the edges $(n_{l_i}, n_{l_{i+1}}) \in E$ it follows that $\mathbb{P}[X_1 = v_{l_{i+1}} | X_0 = v_{l_i}] > 0$. Thus

$$\mathbb{P}[X_n = t | X_0 = s] \geq \prod_{i=0}^{k-1} \underbrace{\mathbb{P}[X_1 = v_{l_{i+1}} | X_0 = v_{l_i}]}_{>0} > 0. \tag{5.22}$$

The first inequality occurs as there can exist other walks from s to t in altogether n steps. For the third case, $s = t$ and we have a connected graph with at least two nodes and therefore non-empty edge sets and node sets. Hence, we can apply Lemma 5.22. □

Remark 5.23. *In general the converse statement of Theorem 5.20 is not true. Irreducible Markov chains can also be induced by two disconnected graphs G_1 and G_2 if at least one node value of G_1 coincides with one node value of G_2.*

The induced Markov chain in the following example is not ergodic.

Model problem 2. Let a graph G like in Figure 5.13 be given.

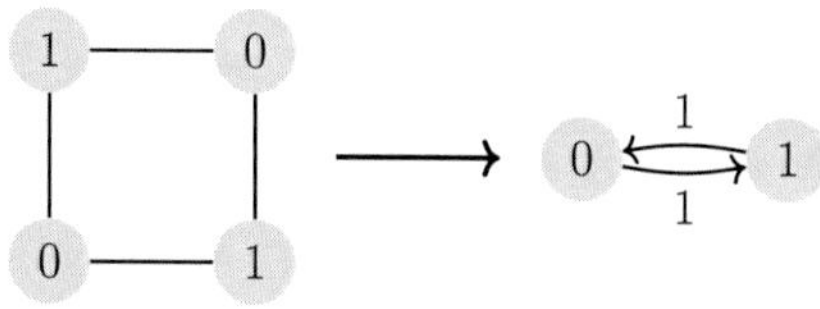

Figure 5.13: *Mean ergodic but not ergodic Markov chain induced by the graph to the left.*

The transition probabilities are

$$\begin{aligned}\mathbb{P}[X_1 = 0 | X_0 = 0] = \mathbb{P}[X_1 = 1 | X_0 = 1] &= 0, \text{ and} \\ \mathbb{P}[X_1 = 1 | X_0 = 0] = \mathbb{P}[X_1 = 0 | X_0 = 1] &= 1.\end{aligned} \tag{5.23}$$

If we start a Markov chain with these transition probabilities and starting state $X_0 = 0$ we get

$$\mathbb{P}[X_i = 0 | X_0 = 0] = \begin{cases} 1 & \text{if } i \in \{2k | k \in \mathbb{N}\} \\ 0 & \text{else} \end{cases}$$

The periodicity for state $X_0 = 0$ of the Markov chain defined by the transition probabilities in Equation (5.23) is 2, see also Equation (3.23). Hence, a necessary condition of the ergodicity is violated. The same statement holds for $X_0 = 1$.

Recalling Remark 3.46, that the convergence speed from an initial distribution to a stationary distribution is dependent on the periodicity and that we calculate the entropy based on a stationary distribution, ergodicity is an important property for our induced Markov chains.

Theorem 5.24. *Let M be a finite state Markov chain induced by a connected and undirected graph $G = (V, E)$ with at least two nodes. The Markov chain M is aperiodic if there exists a walk on G from one node $n_i \in V$ back to itself with an odd length.*

PROOF: Let $G = (V, E)$ be a graph with a walk $\langle n_i, n_{l_1}, ..., n_{l_{k-1}}, n_i \rangle$ of odd length $k \in \{2k - 1 | k \in \mathbb{N}\}$. Furthermore, let the values defined on the node n_i be v_i for $i = 1, ..., n$ and $\mathcal{A} = \{v_i : i \in \{1, ..., n\}\}$.
As we consider relative frequencies and finite graphs the transition probability of two values of neighbored nodes is greater than zero. Thus, if for every node $n_j \in V$ exists a walk

$$\langle n_j, n_{m_1}, ..., n_{m_r}, n_i, n_{l_2}, ..., n_{l_{k-1}}, n_i, n_{m_r}, ..., n_{m_1}, n_j \rangle \tag{5.24}$$

on the graph with odd length from node n_j to node n_i, the known odd length walk back to node n_i and finally the same way back from n_i to n_j. This walk has length $2(r + 1) + k \in \{2k - 1 | k \in \mathbb{N}\}$.
From Lemma 5.22 we know that a walk with even length from n_j to one of its neighbors exists. Thus we can construct for every node $n_j \in V$ walks with length $\{k + 2(r + l) | l \in \mathbb{N}\} \subset \{2k - 1 | k \in \mathbb{N}\}$, where k is the length of the known odd walk from n_i back to itself and r the length of a walk from n_j to n_i. So the set

$$S = \{t \in \mathbb{N} : \mathbb{P}[X_t = v_j | X_0 = v_j] > 0\} \tag{5.25}$$

consists of at least all odd numbers bigger than $k + 2(r+1)$ and therefore especially contains prime numbers. As there are infinite many prime numbers bigger than $k + 2(r + 1)$ the greatest common divisor of S is one and the state $v_j \in \mathcal{A}$ is aperiodic. As n_j was chosen arbitrarily it is valid for all states and thus the Markov chain is aperiodic.   □

Proposition 5.25. *Let $G = (V, E)$ be a graph. If a walk with odd length from one node back to itself exists the induced Markov chain is ergodic.*

With Proposition 5.25, we can independently of node values assure ergodicity based on the topology of the underlying graph. If there is no such walk with odd length the ergodicity can additionally be fulfilled by positions of the node values, see Theorem 5.26.

PROOF: Based on Corollary 3.41 a finite state space, irreducible, and aperiodic Markov chain is ergodic. Since V is finite we have a finite state space. Together with Theorems 5.20 and 5.24 we get the ergodicity of the induced Markov chain.   □

The Model Problem 1 has obviously no walks from one node back to itself with odd length. Nevertheless, the Model Problem 1 is aperiodic due to the connection of two nodes with the same value. Hence, together with the irreducibility, the induced Markov chain is ergodic.

Theorem 5.26. *Let a connected and undirected graph $G = (V, E)$ with at least two nodes and node values $v_1, ..., v_n \in \mathcal{A}$ be given. The induced Markov chain is aperiodic if there* *exists for at least one value $a \in \mathcal{A} = \{v_1, ..., v_n\}$ a walk with odd length from node $n_i \in V$ with value $v_i = a$ to node $n_j \in V$ with the same value $v_j = a$.*

To ensure ergodicity of the induced Markov chain it is sufficient to find a walk with odd length from a node $n_i \in N$ with a value $a \in \mathcal{A}$ to a node n_j with node value a where $n_i = n_j$ is allowed.

PROOF: If there exists an odd walk from one node back to itself in the graph we know by Theorem 5.24 that the induced Markov chain is aperiodic. We now consider the case of walks with even length from one node back to itself.

According to our premises let $a \in \mathcal{A}$ be the node value which occurs at the nodes n_i and n_j together with a walk with odd length

$$\langle n_i, n_{l_1}, ..., n_{l_k}, n_j \rangle$$

with $k \in \{2i | i \in \mathbb{N}\}$ and

$$\langle n_i, n_j \rangle$$

if the two nodes with value $v_i = v_j = a \in \mathcal{A}$ are connected, i.e. $(n_i, n_j) \in E$, respectively. The state $a \in \mathcal{A}$ is aperiodic as we can add to the given walk with odd length walks from n_j to one of its neighbors and back to itself, see Lemma 5.22. So there exist walks of odd length $k + 1 + j$ with $k, j \in \{2i | i \in \mathbb{N}\}$ and so the greatest common divisor of the length of all possible walks is one and thus $a \in \mathcal{A}$ is aperiodic.

Since we have to prove aperiodicity for all elements from the alphabet $\mathcal{A}$, let $b \in \mathcal{A}$ with $b \neq a$ and n_b a node with node value b. We take an arbitrary walk from node n_b to node n_i and a walk from node n_j to node n_b. One of these two walks has to have even length and the other one odd length, compare Proposition A.13,

$$\langle n_b, n_{m_1}, ..., n_{m_s}, n_i \rangle \text{ and } \langle n_j, n_{l_1}, ..., n_{l_t}, n_b \rangle$$

with $s + t \in \{2i - 1 | i \in \mathbb{N}\}$.
Hence,

$$\mathbb{P}[X_n = b | X_0 = b]$$
$$\geq \mathbb{P}[v_{m_1} | v_c] \mathbb{P}[v_i | v_{m_s}] \mathbb{P}[v_c] \prod_{i=1}^{s-1} \mathbb{P}[v_{m_{i+1}} | v_{m_i}] \cdot \mathbb{P}[v_{l_1} | v_j] \mathbb{P}[v_c | v_{l_t}] \mathbb{P}[v_j] \prod_{i=1}^{t-1} \mathbb{P}[v_{l_{i+1}} | v_{l_i}]$$
$$> 0,$$

with $n \in \{2i + 1 > s + t + 1 | i \in \mathbb{N}\}$. Again walks of length two from node n_b to one of its neighbors and back to n_b can be added, see Lemma 5.22.

Thus the greatest common divisor of all n with $\mathbb{P}[X_n = b | X_0 = b] > 0$ is one and the state $b \in \mathcal{A}$ is aperiodic. So every state in $\mathcal{A}$ is aperiodic and thus the Markov chain is aperiodic. □

5.5   Divergence of empirical distribution and stationary distribution of induced Markov chain

In this section, we investigate how well the induced Markov chain model fits to a given data set. For this purpose we investigate the distinction between the stationary distribution

$\mathbb{P}_Z$ of the induced Markov chain, see Theorem 5.17, and the empirical distribution $\mathbb{P}_Y$, see Definition 5.15. Since we assume that the node values are a result of a random walk on the graph setting node values based on the transition probabilities, see Section 5.4, the difference between these two distributions tells whether this assumption is justified. As the central result of this section we determine an upper bound of the divergence solely based on the topology of the graph and independent of alphabets and node value vectors. We use the stationary distribution as representation of the induced Markov chain, since we can compare it with the empirical distribution applying the Kullback-Leibler-Divergence.

Definition 5.27 (Kullback-Leibler-Divergence [25]). *Let $\mathbb{P}_Y$ and $\mathbb{P}_Z$ be two probability mass functions for two random variables Y and Z with the same state space $\mathcal{A}$ and $\mathbb{P}_Z[a] \neq 0$ for all $a \in \mathcal{A}$. We call*

$$KL(\mathbb{P}_Y, \mathbb{P}_Z) = \sum_{a \in \mathcal{A}} \mathbb{P}_Y[a] \log_2 \left(\frac{\mathbb{P}_Y[a]}{\mathbb{P}_Z[a]} \right). \tag{5.26}$$

Kullback-Leibler-Divergence (KLD).

The Kullback-Leibler-Divergence is also known as relative entropy, since it can be interpreted as a measure of additional bits per symbol required to encode values with a distribution $\mathbb{P}_Y$ but using a code based on $\mathbb{P}_Z$ [25].

Lemma 5.28 ([25]). *The Kullback-Leibler-Divergence is nonnegative.*

Since some elements of the sum of Formula 5.26 can be negative it is important to state that the KLD is nonnegative for an application as a distance measure.
In the case of induced Markov chains we can interpret $\mathbb{P}_Y$, see Equation (5.12), as the empirical distribution and $\mathbb{P}_Z$, see Equation (5.18), as the stationary distribution for a random walk on a graph where we set node values from $\mathcal{A}$. If the distributions are very similar, the transition probabilities will depict the distribution of the underlying graph and node value vector very well. Due to the definitions of $\mathbb{P}_Y$ and $\mathbb{P}_Z$, we know that for all $a \in \mathcal{A}$ it holds $\mathbb{P}_Z[a] \neq 0$ if $\mathbb{P}_Y[a] \neq 0$. Together with the minimality of alphabet $\mathcal{A}$, see Remark 5.13, we can ensure $\mathbb{P}_Z[a] \neq 0$ for all $a \in \mathcal{A}$.

Definition 5.29 ([32]). *Let $G = (N, E)$ be a graph. The graph G is called k-regular, if the degree of all nodes is equal to k.*

An example for a k-regular graph is a complete graph with k nodes.

Theorem 5.30. *Let $G = (N, E)$ be a k-regular graph with $n \in \mathbb{N} \backslash \{0\}$ nodes, $e \in \mathbb{N} \backslash \{0\}$ edges, $\mathcal{A} = \{a_1, ..., a_m\}$ an alphabet, and $v \in \mathcal{A}^n$ a node value vector. Then it holds*

$$KL(\mathbb{P}_Y, \mathbb{P}_Z) = 0$$

The KLD is zero independent of the node value vector $v \in \mathcal{A}^n$, because we only use topological properties for our Theorem.

PROOF: From Equation (5.12), (5.18) and $Y(n_i) = v_i$ for all $i = 1, ..., n$, it holds for all $a \in \mathcal{A}$ $\mathbb{P}_Y[a] = \frac{1}{n} \#\{n_i \in N : Y(n_i) = a\}$ and $\mathbb{P}_Z[a] = \frac{1}{e} \#\{(i, \cdot) \in E | v_i = a\}$. Since k

edges start at every node, we have $\#\{(i, \cdot) \in E | Y(n_i) = a\} = k\#\{n_i \in N | Y(n_i) = a\}$ for all $a \in \mathcal{A}$ and $e = k \cdot n$.

$$
\begin{aligned}
\mathbb{P}_Y[a] &= \frac{1}{n}\#\{n_i \in N : Y(n_i) = a\} \\
&= \frac{1}{n}\frac{1}{k}\#\{(i, \cdot) \in E | Y(n_i) = a\} \\
&= \frac{1}{e}\#\{(i, \cdot) \in E | Y(n_i) = a\} \\
&= \mathbb{P}_Z[a] \quad \forall a \in \mathcal{A}
\end{aligned} \tag{5.27}
$$

When we apply Equation (5.27) to the definition of the Kullback Leibler Divergence we get:

$$
\begin{aligned}
KL(\mathbb{P}_Y, \mathbb{P}_Z) &= \sum_{a \in \mathcal{A}} \mathbb{P}_Y[a] \log_2\left(\frac{\mathbb{P}_Y[a]}{\mathbb{P}_Z[a]}\right) \\
&\overset{(5.27)}{=} \sum_{a \in \mathcal{A}} \mathbb{P}_Y[a] \log_2\left(\frac{\mathbb{P}_Y[a]}{\mathbb{P}_Y[a]}\right) \\
&= \sum_{a \in \mathcal{A}} \mathbb{P}_Y[a] \log_2(1) \\
&= 0
\end{aligned}
$$

$\square$

In our application field, k-regular graphs are rare. In the remainder of this section, we, therefore, focus on wider class of graphs namely finite connected graphs.

In the following, we give an upper bound for the Kullback-Leibler-Divergence between the empirical distribution and the stationary distribution of an induced Markov chain for a value vector $v \in \mathcal{A}^n$ with $n = \#N \geq 2$ and a finite alphabet $\mathcal{A}$ and a connected graph $G = (N, E)$. This bound depends solely on the topology of the graph with the given bound being valid for all node value vectors $v \in \mathcal{A}^n$ with $n = \#N$ and a minimal alphabet $\mathcal{A}$, i.e. every symbol $a \in \mathcal{A}$ occurs in the node value vector, see Remark 5.13. We start with the statement that the KLD is independent of the number of symbols in the alphabet, if the node value vector has a special property, namely, is separated by node degree, see Definition 5.31. Next we show, that the KLD is maximal if the node value vector is separated by node degree. We approximate the Kullback-Leibler-Divergence for the special case that all nodes having the same node degree are assigned with the same value, which results in an upper bound of the KLD for connected graphs with at least two nodes.

We introduce the definition of "separated by node degree" for symbols of the alphabet as well as node value vectors.

Definition 5.31 (separated by node degree). *Let $G = (N, E)$ be a connected graph, $\mathcal{A}$ an alphabet, and $v \in \mathcal{A}^n$ with $n = \#N \geq 2$ a node value vector. We call a symbol $a \in \mathcal{A}$ separated by node degree if and only if for every $a \in \mathcal{A}$ there exists a $c_a \in \mathbb{N}\backslash\{0\}$ such that $\deg(n^a) = c_a$ for all $n^a \in \{n_i \in N | v_i = a\}$. We call a node value vector v separated by node degree if and only if all symbols $a \in \mathcal{A}$ are separated by node degree.*

In Figure 5.14, we have a graph of Model Problem 1, a node value vector $v_i = \deg(n_i)$ for $i = 1, ..., 9$ and a minimal alphabet $\mathcal{A} = \{2, 3, 4\}$. This node value vector is obviously separated by node degree.

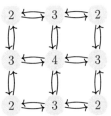

Figure 5.14: *Graph of Model Problem 1 and a node value vector for alphabet $\mathcal{A} = \{2, 3, 4\}$ that is separated by node degree, since for this special case it holds that $v_i = \deg(n_i)$.*

The node value vector of Model Problem 1 is not separated by node degree, since e.g. node value 4 occurs on a node with a degree equal to 2 and on a node with a degree equal to 3, see Figure 5.15. Therefore, the symbol $4 \in \mathcal{A} = \{0, 1, 2, 3, 4, 5\}$ is not separated by node degree and consequently, v is not separated by node degree.

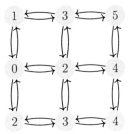

Figure 5.15: *The node value vector for Model Problem 1 is not separated by node degree since e.g. $a = 2$ occurs on a node with degree equal to two and on one node with degree equal to 4.*

In the next step we show, that the KLD coincides for all node value vectors that are separated by node degree on a connected graph G. The following theorem states that the Kullback-Leibler-Divergence for the empirical distribution $\mathbb{P}_Y^v$ and the stationary distribution of the induced Markov chain $\mathbb{P}_Z^v$ is independent of the size of an minimal alphabet, if the node value vector v is separated by node degree.

Theorem 5.32. *Let $G = (N, E)$ be a connected graph with $n \in \backslash\{0\}$ nodes and $e \in \backslash\{0\}$ edges and a node value vector $v \in \mathcal{A}^n$, which is separated by node degree, while $d_1, ..., d_k$ are the k different degrees of graph G. Moreover, let $\tilde{\mathcal{A}} = \{\tilde{a}_1, ..., \tilde{a}_k\}$ be an alphabet, and for all $i = 1, ..., n$, we define*

$$\tilde{v}_i := \tilde{a}_l \text{ if } \deg(n_i) = d_l. \tag{5.28}$$

The vector $\tilde{v}$ is a node value vector with all nodes with one degree having the same value. Let $\mathbb{P}_Y^v$ and $\mathbb{P}_Y^{\tilde{v}}$ be the empirical distribution induced by node value vector $v \in \mathcal{A}^n$ and $\tilde{v} \in \tilde{\mathcal{A}}^n$, respectively. Analogously, let $\mathbb{P}_Z^v$ and $\mathbb{P}_Z^{\tilde{v}}$ be the stationary distribution of the induced Markov chain induced by node value vector $v \in \mathcal{A}^n$ and $\tilde{v} \in \tilde{\mathcal{A}}^n$, respectively. It holds that

$$KL(\mathbb{P}_Y^v, \mathbb{P}_Z^v) = KL(\mathbb{P}_Y^{\tilde{v}}, \mathbb{P}_Z^{\tilde{v}}). \tag{5.29}$$

PROOF: Let $v \in \mathcal{A}^n$ be a node value vector and $a_1^l, ..., a_p^l \in \mathcal{A}$ be the values that occur on nodes with degree d_l for all $l = 1, ..., k$. Let $k_{a_j^l} := \#\{n_i \in N | v_i = a_j^l\} \in \mathbb{N}\backslash\{0\}$ for $j = 1, ..., p$ the number of elements of v that have value a_j^l, while the node has degree d_l. It holds that

$$\tilde{k}_{\tilde{a}_l} := \#\{n_i \in N | \tilde{v}_i = \tilde{a}_l\}$$
$$= \sum_{j=1}^{p} \#\{n_i \in N | v_i = a_j^l\}$$
$$= \sum_{j=1}^{p} k_{a_j^l} \in \mathbb{N}\backslash\{0\}$$

for $l = 1, ..., k$. Then again for all $l = 1, ..., k$, it holds that

$$\sum_{j=1}^{p} \mathbb{P}_Y^v[a_j^l] \log_2\left(\frac{\mathbb{P}_Y^v[a_j^l]}{\mathbb{P}_Z^v[a_j^l]}\right) = \sum_{j=1}^{p} \frac{\#\{n_i \in N | v_i = a_j^l\}}{n} \log_2\left(\frac{\frac{\#\{n_i \in N | v_i = a_j^l\}}{n}}{\frac{\#\{(i,\cdot) \in E | v_i = a_j^l\}}{e}}\right)$$
$$= \sum_{j=1}^{p} \frac{k_{a_j^l}}{n} \log_2\left(\frac{\frac{k_{a_j^l}}{n}}{\frac{d_l k_{a_j^l}}{e}}\right)$$
$$= \sum_{j=1}^{p} \frac{k_{a_j^l}}{n} \log_2\left(\frac{e}{d_l n}\right)$$
$$= \tilde{k}_{\tilde{a}_l} \log_2\left(\frac{e}{d_l n}\right)$$
$$= \tilde{k}_{\tilde{a}_l} \log_2\left(\frac{\frac{\tilde{k}_{\tilde{a}_l}}{n}}{\frac{d_l \tilde{k}_{\tilde{a}_l}}{e}}\right)$$
$$= \frac{\#\{n_i \in N | \tilde{v}_i = \tilde{a}_l\}}{n} \log_2\left(\frac{\frac{\#\{n_i \in N | \tilde{v}_i = \tilde{a}_l\}}{n}}{\frac{\#\{(i,\cdot) \in E | \tilde{v}_i = \tilde{a}_l\}}{e}}\right)$$
$$= \mathbb{P}_Y^{\tilde{v}}[\tilde{a}_l] \log_2\left(\frac{\mathbb{P}_Y^{\tilde{v}}[\tilde{a}_l]}{\mathbb{P}_Z^{\tilde{v}}[\tilde{a}_l]}\right). \tag{5.30}$$

These equations hold for all node degrees $d_1, ..., d_k$ and since v is separated by node degrees, we get

$$KL(\mathbb{P}_Y^v, \mathbb{P}_Z^v) = \sum_{l=1}^{k} \sum_{a \in \{a_j^l \in \mathcal{A} | j=1,...,p\}} \mathbb{P}_Y^v[a] \log_2\left(\frac{\mathbb{P}_Y^v[a]}{\mathbb{P}_Z^v[a]}\right)$$

$$\overset{(5.30)}{=} \sum_{l=1}^{k} \mathbb{P}_Y^{\tilde{v}}[\tilde{a}_l] \log_2 \left(\frac{\mathbb{P}_Y^{\tilde{v}}[\tilde{a}_l]}{\mathbb{P}_Z^{\tilde{v}}[\tilde{a}_l]} \right)$$

$$= KL(\mathbb{P}_Y^{\tilde{v}}, \mathbb{P}_Z^{\tilde{v}})$$

□

As a consequence of Theorem 5.32 the entropy of node value vector is effected by the size of the alphabet, see Formula (3.24), but the relative entropy, i. e. the KLD, is independent. Therefore, the model of induced Markov chains fits to the relative frequency model in both big and small alphabets.

In the following we show, that for a given node value vector $v \in \mathcal{A}^n$, which is not separated by node degree, there exists a node value vector $\tilde{v}$ separated by node degree, and its KLD is at least as big as the one induced by v. We start with the observation for one symbol that is not separated by node degree.

Lemma 5.33. *Let $G = (N, E)$ be a connected graph with $n \in \mathbb{N}$, $n \geq 2$, nodes, $N = \{n_1, ..., n_n\}$ and $e \in \mathbb{N}$, $e \geq 1$, edges, $E = \{e_1, ...e_e\}$, $\mathcal{A} = \{a_1, ..., a_m\}$ an alphabet, $v \in \mathcal{A}^n$ a node value vector and $b \in \mathcal{A}$ a symbol that occurs on nodes $n_1^b, ..., n_l^b$, $l \in \mathbb{N}\backslash\{0\}$ with node degrees $d_1, ..., d_k \in \mathbb{N}\backslash\{0\}$, i.e. $\deg(n_i^b) \in \mathcal{D} = \{d_1, ..., d_k\}$ for all $i = 1, ..., l$. Moreover, let $\mathcal{D}$ be minimal in respect to its number of elements, i.e. $\exists n_i^b : \deg(n_i^b) = d_j$ for all $j = 1, ..., k$. For all $j = 1, .., k$, we define*

$$k_j^b := \#\{n_i^b \in N | \deg(n_i^b) = d_j\}.$$ (5.31)

Then, for all $b \in \mathcal{A}$, it holds that

$$\frac{\sum_{j=1}^{k} k_j^b}{n} \log_2 \left(\frac{\frac{\sum_{j=1}^{k} k_j^b}{n}}{\frac{\sum_{j=1}^{k} k_j^b d_j}{e}} \right) \leq \sum_{j=1}^{k} \frac{k_j^b}{n} \log_2 \left(\frac{\frac{k_j^b}{n}}{\frac{k_j^b d_j}{e}} \right).$$ (5.32)

For the proof of Lemma 5.33 we need an approximation of the logarithm.

Lemma 5.34 ([105]). *For $x \in \mathbb{R}$ and $x > 1$, it holds for the natural logarithm that*

$$1 - \frac{1}{x} < \ln(x) < x - 1$$

From Lemma 5.34, we can follow that:

$$\frac{1 - \frac{1}{x}}{\ln(2)} < \frac{\ln(x)}{\ln(2)} = log_2(x) < \frac{x - 1}{\ln(2)}, \forall x \in \mathbb{R} \wedge x > 1$$ (5.33)

In the following, we give the proof of Lemma 5.33.

PROOF: Let $N = \{n_1, ..., n_n\}$ be a node set with $n \geq 2$ and $E = \{e_1, ..., e_e\}$, the edge set with $e \geq 1$. Due to the definition of the node degree set $\mathcal{D}$, we have $k_j^b \geq 1$ for all $j = 1, ..., k$. Moreover, $d_i > 0$ for all $i = 1, ..., k$ since G is connected. Let $A_i^b := \frac{\sum_{j=1}^{k} k_j^b d_j}{d_i \sum_{j=1}^{k} k_j^b}$ for all $i = 1, ..., k$. Since all elements of A_i^b are positive it holds that $A_i^b > 0$. We split the set $\{1, ..., k\}$ into the following three sets:

$$\mathcal{I}_>^b := \{i \in \{1, ..., k\} | A_i^b > 1\}$$

$$\mathcal{I}^b_= := \{i \in \{1, ..., k\}|A^b_i = 1\}$$
$$\mathcal{I}^b_< := \{i \in \{1, ..., k\}|A^b_i < 1\}.$$

Since A^b_i is positive we also get that

$$\log_2(A^b_i) > 0, \forall i \in \mathcal{I}^b_>$$
$$\log_2(A^b_i) = 0, \forall i \in \mathcal{I}^b_=$$
$$\log_2(A^b_i) < 0, \forall i \in \mathcal{I}^b_<$$

The next set of equations shows, that the right hand side of Equation (5.32) minus its left hand side is nonnegative, which is equivalent to Equation (5.32):

$$0 = \frac{1}{n \ln(2)} \left(\sum_{i=1}^{k} k^b_i - \left(\sum_{j=1}^{k} k^b_j \right) \frac{\sum_{i=1}^{k} k^b_i d_i}{\sum_{j=1}^{k} k^b_j d_j} \right)$$

$$= \frac{1}{n \ln(2)} \left(\sum_{i=1}^{k} k^b_i - \frac{\sum_{i=1}^{k} k^b_i d_i \left(\sum_{j=1}^{k} k^b_j \right)}{\sum_{j=1}^{k} k^b_j d_j} \right)$$

$$= \frac{1}{n \ln(2)} \left(\sum_{i=1}^{k} k^b_i \left(1 - \frac{d_i \sum_{j=1}^{k} k^b_j}{\sum_{j=1}^{k} k^b_j d_j} \right) \right)$$

$$= \frac{1}{n \ln(2)} \left(\sum_{i=1}^{k} k^b_i \left(1 - \left(A^b_i \right)^{-1} \right) \right)$$

$$= \frac{1}{n \ln(2)} \left(\underbrace{\sum_{i \in \mathcal{I}^b_>} k^b_i \underbrace{\left(1 - \frac{1}{A^b_i} \right)}_{>0}}_{>0} + \underbrace{\sum_{i \in \mathcal{I}^b_=} k^b_i \underbrace{\left(1 - \left(A^b_i \right)^{-1} \right)}_{=0}}_{=0} - \underbrace{\sum_{i \in \mathcal{I}^b_<} k^b_i \underbrace{\left(\left(A^b_i \right)^{-1} - 1 \right)}_{>0}}_{<0} \right).$$

When we apply the first estimation from (5.33) to all terms of the first sum and and the second estimation from (5.33) on all terms of the third sum, we get the following estimation:

$$0 \overset{(5.33)}{\le} \frac{1}{n \ln(2)} \left(\sum_{i \in \mathcal{I}^b_>} k^b_i \log_2 A^b_i \ln(2) + \underbrace{\sum_{i \in \mathcal{I}^b_=} k^b_i \log_2 A^b_i \ln(2)}_{=0} - \sum_{i \in \mathcal{I}^b_<} k^b_i \log_2 \left(A^b_i \right)^{-1} \ln(2) \right)$$

$$= \frac{1}{n} \left(\sum_{i \in \mathcal{I}^b_>} k^b_i \log_2 A^b_i + \sum_{i \in \mathcal{I}^b_=} k^b_i \log_2 A^b_i + \sum_{i \in \mathcal{I}^b_<} k^b_i \log_2 A^b_i \right)$$

$$= \frac{1}{n} \left(\sum_{i=1}^{k} k^b_i \log_2 A^b_i \right)$$

$$= \frac{1}{n} \left(\sum_{i=1}^{k} k^b_i \log_2 \left(\frac{\sum_{j=1}^{k} k^b_j d_j}{d_i \sum_{j=1}^{k} k^b_j} \right) \right)$$

$$= \frac{1}{n} \left(\sum_{i=1}^{k} k^b_i \left(\log_2 \left(\frac{\sum_{j=1}^{k} k^b_j d_j}{\sum_{j=1}^{k} k^b_j} \right) - \log_2(d_i) + \underbrace{\log_2 \left(\frac{n}{e} \right) - \log_2 \left(\frac{n}{e} \right)}_{=0} \right) \right)$$

$$= \frac{1}{n}\left(\sum_{i=1}^{k} k_i^b \left(\log_2\left(\frac{\sum_{j=1}^{k} k_j^b d_j}{\sum_{j=1}^{k} k_j^b}\right) + \log_2\left(\frac{n}{e}\right)\right) - \sum_{i=1}^{k} k_i^b\left(\log_2(d_i) + \log_2\left(\frac{n}{e}\right)\right)\right)$$

$$= \frac{1}{n}\left(\sum_{i=1}^{k} k_i^b\left(\log_2\left(\frac{\frac{\sum_{j=1}^{k} k_j^b d_j}{e}}{\frac{\sum_{j=1}^{k} k_j^b}{n}}\right)\right) - \sum_{i=1}^{k} k_i^b\left(\log_2\left(\frac{\frac{k_i^b d_i}{e}}{\frac{k_i^b}{n}}\right)\right)\right)$$

$$= \sum_{i=1}^{k} \frac{k_i^b}{n}\left(\log_2\left(\frac{\frac{k_i^b}{n}}{\frac{k_i^b d_i}{e}}\right)\right) - \frac{\sum_{i=1}^{k} k_i^b}{n}\left(\log_2\left(\frac{\frac{\sum_{j=1}^{k} k_j^b}{n}}{\frac{\sum_{j=1}^{k} k_j^b d_j}{e}}\right)\right).$$

$\square$

Lemma 5.33 allows to find an upper bound for each element that occur in the formula of the KLD applied on $\mathbb{P}_Y$ and $\mathbb{P}_Z$.

Theorem 5.35. *Let* $G = (N,E)$ *be a connected graph with* $n \in \mathbb{N}$, $n \geq 2$, *nodes,* $N = \{n_1, ..., n_n\}$ *and* $e \in \mathbb{N}$, $e \geq 1$, *edges,* $E = \{e_1, ...e_e\}$, $\mathcal{A} = \{a_1, ..., a_m\}$ *a minimal alphabet, and a node value vector* $v \in \mathcal{A}^n$. *Then, there exists a minimal alphabet* $\tilde{\mathcal{A}}$ *and a node value vector* $\tilde{v} \in \tilde{\mathcal{A}}^n$ *that is separated by node degree such that*

$$KL(\mathbb{P}_Y^v, \mathbb{P}_Z^v) \leq KL(\mathbb{P}_Y^{\tilde{v}}, \mathbb{P}_Z^{\tilde{v}}). \tag{5.34}$$

PROOF: Let a node value vector $v \in \mathcal{A}^n$ be given. If v is separated by node degree, we set $\tilde{v} = v$, $\tilde{\mathcal{A}} = \mathcal{A}$ and Equation (5.34) is trivially fulfilled. If v is not separated by node degree, there exists symbols $b_1, ..., b_s \in \mathcal{A}$ with $s \in \mathbb{N}\backslash\{0\}$, that are not separated by node degree, which occurs on nodes $N^{b_j} = \{n_i \in N | v_i = b_j\}$ with degrees $d_1^{b_j}, ..., d_{k_j}^{b_j}$ for $j = 1, ..., s$. Moreover, let $a_{m+1}^{(j)}, ..., a_{m+k_j-1}^{(j)} \notin \mathcal{A}$ and pairwise distinct for all $j = 1, ..., s$. We define for all $i = 1, ..., n$

$$\tilde{v}_i = \begin{cases} v_i & n_i \in N^{b_j} \wedge \deg(n_i) = d_1^{b_j}, j \in \{1, ..., s\} \\ a_{m+l}^{(j)} & n_i \in N^{b_j} \wedge \deg(n_i) = d_{l+1}^{b_j}, l \in \{1, ..., k-1\}, j \in \{1, ..., s\} \\ v_i & \text{others.} \end{cases}$$

and therefore, $\tilde{v} \in \tilde{\mathcal{A}} := \bigcup_{j=1}^{s}\{a_{m+1}^{(j)}, ..., a_{m+k_j-1}^{(j)}\} \cup \{a_1, ..., a_m\}$. The node value vector $\tilde{v}$ is separated by node degree, since we give for every degree $d_i^{b_j}$ with $i \geq 2$, $j \in \{1, ..., s\}$ a new symbol. Let $k_j^b := \#\{n_i^b \in N | \deg(n_i^b) = d_j\}$ for all $b \in \{b_1, ..., b_s\}$, see Equation (5.31).

$$KL(\mathbb{P}_Y^v, \mathbb{P}_Z^v) = \sum_{a \in \mathcal{A}} \mathbb{P}_Y^v[a] \log_2\left(\frac{\mathbb{P}_Y^v[a]}{\mathbb{P}_Z^v[a]}\right)$$

$$= \sum_{a \in \mathcal{A}\backslash\{b_1, ..., b_s\}} \mathbb{P}_Y^v[a] \log_2\left(\frac{\mathbb{P}_Y^v[a]}{\mathbb{P}_Z^v[a]}\right) + \sum_{a \in \{b_1, ..., b_s\}} \mathbb{P}_Y^v[a] \log_2\left(\frac{\mathbb{P}_Y^v[a]}{\mathbb{P}_Z^v[a]}\right)$$

$$\stackrel{(5.30)}{=} \sum_{a \in \mathcal{A}\backslash\{b_1, ..., b_s\}} \mathbb{P}_Y^{\tilde{v}}[a] \log_2\left(\frac{\mathbb{P}_Y^{\tilde{v}}[a]}{\mathbb{P}_Z^{\tilde{v}}[a]}\right) + \sum_{b \in \{b_1, ..., b_s\}} \mathbb{P}_Y^v[b] \log_2\left(\frac{\mathbb{P}_Y^v[b]}{\mathbb{P}_Z^v[b]}\right)$$

$$= \sum_{a \in \mathcal{A}\backslash\{b_1, ..., b_s\}} \mathbb{P}_Y^{\tilde{v}}[a] \log_2\left(\frac{\mathbb{P}_Y^{\tilde{v}}[a]}{\mathbb{P}_Z^{\tilde{v}}[a]}\right) + \sum_{j=1}^{s} \frac{\sum_{l=1}^{k_j} k_l^{b_j}}{n} \log_2\left(\frac{\frac{\sum_{l=1}^{k_j} k_l^{b_j}}{n}}{\frac{\sum_{l=1}^{k_j} k_l^{b_j} d_l^{b_j}}{e}}\right)$$

$$
\begin{aligned}
\overset{(5.32)}{\leq} & \sum_{a\in\mathcal{A}\backslash\{b_1,\ldots,b_s\}} \mathbb{P}_Y^{\tilde{v}}[a] \log_2\left(\frac{\mathbb{P}_Y^{\tilde{v}}[a]}{\mathbb{P}_Z^{\tilde{v}}[a]}\right) + \sum_{j=1}^{s}\sum_{l=1}^{k_j} \frac{k_l^{b_j}}{n} \log_2\left(\frac{k_l^{b_j}}{\frac{n}{k_l^{b_j}d_l}}\right) \\
= & \sum_{a\in\mathcal{A}\backslash\{b_1,\ldots,b_s\}} \mathbb{P}_Y^{\tilde{v}}[a] \log_2\left(\frac{\mathbb{P}_Y^{\tilde{v}}[a]}{\mathbb{P}_Z^{\tilde{v}}[a]}\right) + \sum_{b\in\bigcup_{j=1}^{s}\{a_{m+1}^{(j)},\ldots,a_{m+k_j-1}^{(j)}\}\cup\{b_1,\ldots,b_s\}} \mathbb{P}_Y^{\tilde{v}}[b] \log_2\left(\frac{\mathbb{P}_Y^{\tilde{v}}[b]}{\mathbb{P}_Z^{\tilde{v}}[b]}\right) \\
= & \sum_{a\in\tilde{\mathcal{A}}} \mathbb{P}_Y^{\tilde{v}}[a] \log_2\left(\frac{\mathbb{P}_Y^{\tilde{v}}[a]}{\mathbb{P}_Z^{\tilde{v}}[a]}\right) \\
= & \, KL(\mathbb{P}_Y^{\tilde{v}}, \mathbb{P}_Z^{\tilde{v}})
\end{aligned}
$$

□

We now can find an upper bound for the Kullback-Leibler-Divergence of the empirical distribution of an arbitrary node value vector and the stationary distribution of its induced Markov chain only based on the topology of a connected graph. From Theorem 5.35 we know that we can find for every node value vector v another node value vector $\tilde{v}$ that is separated by node degree, whose KLD is at least as big as the one of v. Additionally, Kullback-Leibler-Divergence coincides for all node value vectors separated by node degree, see Theorem 5.32. We can, therefore, choose an arbitrary node value vector that is separated by node degree and determine its KLD being an upper bound for all node value vectors for that specific graph.

Theorem 5.36. *Let $G = (N, E)$ be a connected graph with $N = \{n_1, \ldots, n_n\}$ and $E = \{e_1, \ldots, e_e\}$. Moreover, let $\mathcal{D} = \{d_1, \ldots, d_k\}$ be the set of node degrees with $\deg(n_i) \in \mathcal{D}$ and $\forall d_j \in \mathcal{D} : \exists n_i : \deg(n_i) = d_j$, i.e. $\mathcal{D}$ is minimal in respect to its number of elements. Furthermore, let*

$$k_j = \#\{n_i \in N|\deg(n_i) = d_j\} \; \forall j = 1, \ldots, k. \tag{5.35}$$

Then it holds that for all minimal alphabets $\mathcal{A}$ and all node value vectors $v \in \mathcal{A}^n$:

$$KL(\mathbb{P}_Y^v, \mathbb{P}_Z^v) \leq \sum_{j=1}^{k} \frac{k_j}{n} \log_2\left(\frac{e}{d_j n}\right) \tag{5.36}$$

PROOF: Let $\hat{\mathcal{A}} = \{\hat{a}_1, \ldots, \hat{a}_k\}$ be a minimal alphabet. We define $\hat{v} \in \hat{\mathcal{A}}^n$ as

$$\hat{v}_i := \hat{a}_j \text{ if } i \in \{i \in \{1, \ldots, n\}|\deg(n_i) = d_j\}$$

for all $i = 1, \ldots, n$ and $j \in \{1, \ldots, k\}$. With the definition of the frequencies of node degrees k_j, see Equation (5.35), we get

$$
\begin{aligned}
KL(\mathbb{P}_Y^{\hat{v}}, \mathbb{P}_Z^{\hat{v}}) &= \sum_{j=1}^{k} \frac{k_j}{n} \log_2\left(\frac{\frac{k_j}{n}}{\frac{k_j d_j}{e}}\right) \\
&= \sum_{j=1}^{k} \frac{k_j}{n} \log_2\left(\frac{e}{d_j n}\right)
\end{aligned}
\tag{5.37}
$$

Let $\mathcal{A}$ be a minimal alphabet and $v \in \mathcal{A}$ a node value vector. From Theorem 5.35 we know that we can find an alphabet $\tilde{\mathcal{A}}$ and a node value vector $\tilde{v} \in \tilde{\mathcal{A}}^n$, that is separated by node degree and fulfills Equation (5.34).

$$KL(\mathbb{P}_Y^v, \mathbb{P}_Z^v) \overset{(5.34)}{\leq} KL(\mathbb{P}_Y^{\hat{v}}, \mathbb{P}_Z^{\hat{v}})$$

$$\overset{(5.29)}{=} KL(\mathbb{P}_Y^{\hat{v}}, \mathbb{P}_Z^{\hat{v}})$$

$$\overset{(5.37)}{=} \sum_{j=1}^{k} \frac{k_j}{n} \log_2 \left(\frac{e}{d_j n} \right)$$

$\square$

Theorem 5.30 is a special case of Theorem 5.36, since for a k-regular graph it holds that $e = k \cdot n$ and therefore $\log_2 \left(\frac{e}{kn} \right) = 0$ and the KLD is nonnegative, see Lemma 5.28.
We can now determine the maximal Kullback-Leibler-Divergence of the empirical distribution and the stationary distribution of the induced Markov chain for all alphabets and the graph G of Model Problem 1, see Figure 5.14. We assume, that the empirical distribution is the correct distribution for that data set. The KLD gives us the number of additional bits, that we need to compress a symbol that has distribution $\mathbb{P}_Y^v$ but we use an optimal code for distribution $\mathbb{P}_Z^v$, see Definition 5.27 and [25]. The stationary distribution reflects the mean distribution of the result of a random walk on G with initial distribution and transition probabilities specified by the induced Markov chain. If the stationary distribution does not match to the empirical distribution, then the underlying node value vector would be a unlikely result of the random walk of the induced Markov chain. In such a case the iMc encoder would gain bad compression rates. The KLD represents these costs in number of bits per symbol. Obviously, it holds that the smaller the KLD is the better the iMc model fits to the data set. For the graph of Model Problem 1 we have the following parameters:

- number of edges of directed graph $e = 24$

- number of nodes $n = 9$

- number of different node degrees $k = 3$

- set of node degrees $\mathcal{D} = \{2, 3, 4\}$

- $k_1 = 4$ nodes with node degree $d_1 = 2$

- $k_2 = 4$ nodes with node degree $d_2 = 3$

- $k_3 = 1$ node with node degree $d_3 = 4$

Let $\mathcal{A}$ be an alphabet and $v \in \mathcal{A}^9$ a node value vector for the graph of Model Problem 1. Let $\mathbb{P}_Y^v$ be the empirical distribution and $\mathbb{P}_Z^v$ the stationary distribution of the induced Markov chain. We can determine an upper bound for $KL(\mathbb{P}_Y^v, \mathbb{P}_Z^v)$ by Equation (5.36) of Theorem 5.36.

$$KL(\mathbb{P}_Y^v, \mathbb{P}_Z^v) \leq \sum_{j=1}^{k} \frac{k_j}{n} \log_2 \left(\frac{e}{d_j n} \right)$$

$$= \frac{4}{9} \log_2 \left(\frac{24}{2 \cdot 9} \right) + \frac{4}{9} \log_2 \left(\frac{24}{3 \cdot 9} \right) + \frac{1}{9} \log_2 \left(\frac{24}{4 \cdot 9} \right)$$

$$= \frac{4}{9} \left(\log_2 \left(\frac{32}{27} \right) \right) + \frac{1}{9} \left(\log_2 \left(\frac{2}{3} \right) \right)$$

$$= \frac{1}{9} \left(\log_2 \left(\frac{2^{20}}{3^{12}} \right) + \log_2 \left(\frac{2}{3} \right) \right)$$

$$= \frac{1}{9} \left(\log_2 \left(\frac{2^{21}}{3^{13}} \right) \right)$$

$$< 0.044$$

The KLD of the empirical distribution $\mathbb{P}_Y^v$ and the stationary distribution of the induced Markov chain $\mathbb{P}_Z^v$ for an arbitrary node value vector and the graph of Model Problem 1 is less than 0.044. Since the iMc model uses information about previously processed values, the overall compression rate is much better compared to the relative frequency coding, see Remark 3.68 and Remark 5.6. For an application on Model Problem 1, we loose less than one bit for the compression of the complete value vector and gain the possibility to exploit dependencies between neighbored values.

5.6   Choice of a spanning tree

In this section we explain briefly, the relations between the selection of a spanning tree and the achieved compression rate. The observed transitions effect the distribution and therefore, the statistics used in an entropy encoder. We tackle the problem briefly only as the overhead of additionally storing the structure of a tree counterbalancing the savings for better encoding. In the case of a set of simulation results, this can be different. For small examples the effect is usually bigger than for graphs with several thousand nodes - this observation might be subject of future works.

The spanning tree from Figure 5.3 and its costs for encoding the transitions for Model problem 1 can be found in Figure 5.16. For an application of adaptive iMc encoding

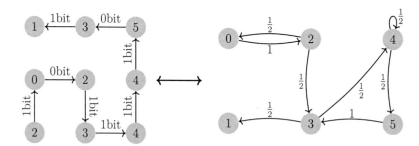

Figure 5.16: Costs for encoding the graph based values with a nonadaptive iMc encoder for the tree of Figure 5.3. The costs are integers since all probabilities are powers of 2. A transition with probability 1 is free.

see Section 5.3, the costs are shown in Figure 5.17. In Figure 5.18, we determined an optimized tree for encoding the graph based data with an adaptive iMc encoder. In this setting, we neglect the size of the statistics, which has to be taken into account for calculating the overall compression factor.

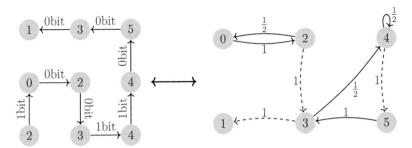

Figure 5.17: *Costs for encoding the graph based values with an adaptive iMc encoder, see Section 5.3, for the tree of Figure 5.3. The costs are integers since all probabilities are integer powers of 2 in all steps of the adaptive encoding. The dashed lines mark transition probabilities that are update in the encoding process.*

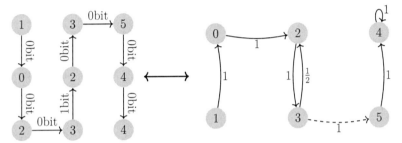

Figure 5.18: *Optimal tree for iMc encoding and its costs for encoding the graph based values with an adaptive iMc encoder. The dashed lines mark transition probabilities that are updated in the encoding process. The number of bits for encoding all transitions is 1.*

Especially with small graphs or badly predicted data sets, the choice of a spanning tree can have a huge influence on the compression rate. The general rule is that the lower the uncertainty for a transition the better it can be compressed by the iMc encoder. Therefore, we use the depth-first search strategy, see [24], to determine a spanning tree, since this approach leads to a high number of different starting nodes for edges that are traversed in the encoding process. An example for a tree that is determined by breadth first search algorithm can be found in Figure 5.19. For this example the overall costs for encoding all transitions is

$$7 = \lceil 6.75 \rceil$$

bits. This is distinctly more than the costs for encoding the transitions for the graph that was found by the depth first search algorithm from the C++ library *boost* [28], see Figure 5.3.

5.7 Combination with prediction methods

In this section, we investigate several prediction methods that can directly be combined with the iMc encoding. We focus on prediction methods that exploit the information

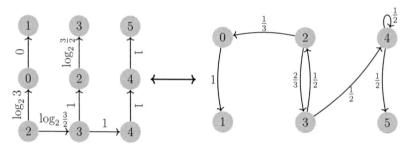

Figure 5.19: Breadth first search determined tree for iMc encoding and its costs for encoding the graph based values with a non-adaptive iMc encoder notated in bits. The overall cost for encoding all transitions will be 7 bits.

about the topology of the mesh and determine a traversal through the graph. The reuse of the traversal is beneficial in the sense of run time and memory consumption. Therefore, it has no drawback compared to encoders that just interpret the data as a one dimensional stream.

Encoding is one of the essential steps of data compression for simulation results. Standard encoding methods such as entropy encoding and dictionary based encoding, see Section 3.3.3, lack the ability to exploit trends in the data. The reduction in disk space is achieved during the encoding step. To improve encoding we can apply bijective prediction methods that modify the distribution in a way that the encoder can benefit of. For the induced Markov chains as well as for all statistical encoders the compression ratio depends on the resulting distribution of a dataset. The better the statistical model fits to the data set to be compressed the better the encoder can reduce the number of bits per symbol. When we apply a prediction method we have to check if the statistical model of the predicted data set still fulfills the required conditions of the encoder.

Remark 5.37. *The better the data set is predicted the more sophisticated the modeling has to be to exploit the remaining dependencies for achieving even better compression factors.*

We distinguish between a practical application of iMc in combination with a prediction method and a theoretical investigation. For the later, we need the entropy to estimate the compression rates and bits per symbol via the well-known theorems of information theory, see Section 3.1. For the practical approach, we determine all transitions frequencies and, therefore, all events to be encoded. Hence, we can determine the compression rate and bits per symbol directly without using the entropy. Nevertheless, the possibility to calculate a well-defined entropy is important, since it reflects the limit for compression under the assumption of a certain distribution. This makes the modeling of a data stream with our induced Markov chain comparable.

We start with practical investigations followed by a theoretical analysis of graph differences. The graph based differences form a superset of graph based prediction methods using direct neighbors for prediction. We close with a brief investigation on compression rates combining prediction and encoding.

5.7.1 Tree differences

In this section, we briefly describe the combination of tree differences, see Section 3.3.2.3, and the induced Markov chain encoder, see Section 5.4. For an example of one and two applications of the tree differences on Model problem 1, see Figure 5.20 and Figure 5.21, respectively. A big advantage of the combination of the tree differences and the iMc

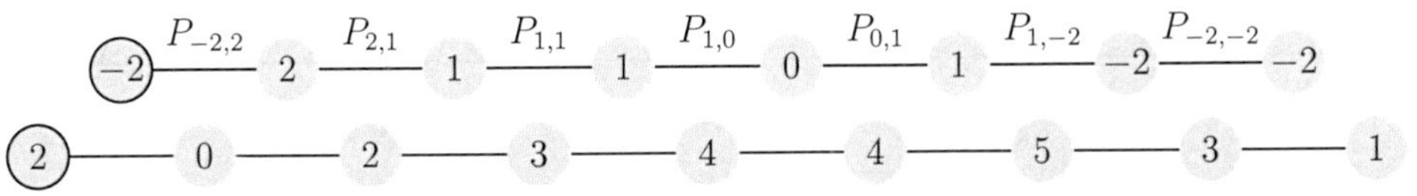

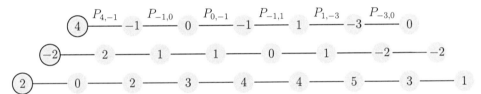

Figure 5.20: *Example from Figure 5.1 with one tree difference. The upper row shows the values after one tree difference. The bold-framed values and the 7 transition probabilities are applied as side information by the iMc encoder.*

Figure 5.21: *Example from Figure 5.1 with two tree differences. The upper row shows the values after two tree differences. The bold-framed values, and the 6 transition probabilities are applied as side information by the iMc encoder.*

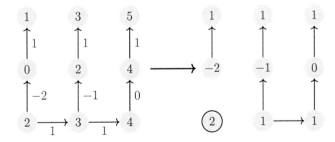

Figure 5.22: *Example from Figure 5.19 with one tree difference.*

encoder is that in the standard formulation of iMc coding we need the tree as a traversal of the graph and vice versa. We, therefore, can use the same tree for prediction and encoding.

Although we use the same tree for both decorrelation and encoding, there are still dependencies left after the prediction that can be exploited in the encoding stage, see [96].

5.7.2   Lorenzo predictor

In this section we introduce our combination of the Lorenzo predictor with the induced Markov chains. For more information on the prediction method and the iMc encoding, we refer to Sections 3.3.2 and 5.1, respectively.

We distinguish between a two and a three dimensional case. In both cases several combinations are possible.

As 2D shell elements are the dominating element type in the simulation of crash tests, see Section 2.2, we start with a two dimensional case. One property of the induced Markov chains is, that they can be defined not only on a tree, but also on directed acyclic graphs (DAG). The parallelogram predictor applies two direct neighbors' values and one value of a neighbor with degree 2 to predict the value of a node. When we determine a strategy to walk through the graph we can exploit the information of the two neighbors that are used for prediction. This leads us to approach one, which uses directed acyclic graphs. This procedure is motivated by the MRF, see Section 3.3.3.7.

We introduce four different approaches that differ in the strategy to set conditions and discuss their advantages and disadvantages.

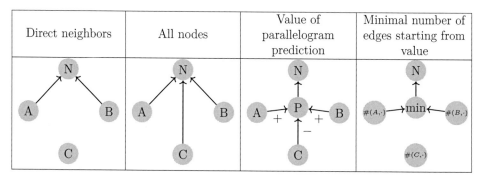

Direct neighbors	All nodes	Value of parallelogram prediction	Minimal number of edges starting from value

Table 5.3: *Four combinations of the information, available for the Lorenzo predictor and the iMc encoder, to determine the statistics for encoding. In the left case only the values of direct neighbors of N are used. In the second case all nodes of the parallelogram are used. In the third case not the values themselves but the combination like in the parallelogram predictor case is applied. In the fourth case the minimum of frequencies for A and B as edge start values is used.*

1. Direct neighbors

2. All nodes used in Lorenzo predictor

3. Value of prediction

4. Minimal number of starting edge values

The investigations for the two dimensional cases can straightforward be generalized for the three dimensional case. Making use of the graph for prediction and encoding enables a big variety of strategies to find a well-fitting statistic. Since we perform an asymmetric compression several distributions can be determined. Since we can determine the compression ratio based on the transition frequencies, we can choose the one that results in the best compression factor.

In first case, where we only consider direct neighbors, we have an inhomogeneous Markov model in the case of adaptive iMc encoding and a homogeneous Markov model if do not use adaptive statistics. In the second to fourth cases we have a Bayesian network model since we consider non-connected nodes for the distribution, see Remark 3.72.

5.7.2.1 Direct neighbors

In the first approach, we only consider the direct neighbors of the node whose value we want to encode.

Let T define the triangle consisting of the direct neighbors n_{i_1}, n_{i_2} of the node n_j with values v_{i_1}, v_{i_2} and v_j, respectively, with $v_{i_1}, v_{i_2}, v_j \in \mathcal{A}$, see also Table 5.3. When we apply a similar strategy like in Section 5.4, we get a distribution that is based on the relative frequencies of the triple (v_{i_1}, v_{i_2}, v_j)

$$\mathbb{P}[N = v_j | A = v_{i_1}, B = v_{i_2}] = \frac{\#\{(n_{i_1}, n_{i_2}, n_j) \in T | v_{i_1} = A, v_{i_2} = B, v_j = N\}}{\#\{(n_{i_1}, n_{i_2}, n_j) \in T | v_{i_1} = A, v_{i_2} = B\}}. \quad (5.38)$$

Since the distribution $\mathbb{P}[N = v_j | A = v_{i_1}, B = v_{i_2}]$ is dependent on two conditions $A = v_{i_1}$ and $B = v_{i_2}$, the transition frequencies form a three dimensional tensor. As long as we assume a dependency between the events $A \cap N$ and $B \cap N$, we cannot simplify the statistic. Using a tensor for managing the distribution has two drawbacks. First the access times of encoding and decoding phase are slower compared to the transition frequency matrix distribution, see Equation 5.2. Second the amount of data that has to be stored is bigger. The challenge is to compensated the overhead for the statistic by better encoding.

5.7.2.2 All nodes used in Lorenzo predictor

The second approach takes all nodes that are used in the Lorenzo predictor to determine a statistic.

Let Q define the quadrangle consisting of the two direct neighbors n_{i_1}, n_{i_2} and a second degree neighbor n_{i_3} of the node n_j with values v_{i_1}, v_{i_2}, v_{i_3} and v_j, respectively, with $v_{i_1}, v_{i_2}, v_{i_3}, v_j \in \mathcal{A}$, see also Table 5.3.

$$\mathbb{P}[N = v_j | A = v_{i_1}, B = v_{i_2}, C = v_{i_3}]$$
$$= \frac{\#\{(n_{i_1}, n_{i_2}, n_{i_3}, n_j) \in Q | v_{i_1} = A, v_{i_2} = B, v_{i_3} = C, v_j = N\}}{\#\{(n_{i_1}, n_{i_2}, n_{i_3}, n_j) \in Q | v_{i_1} = A, v_{i_2} = B, v_{i_2} = C\}}. \quad (5.39)$$

The distribution $\mathbb{P}[N | A, B, C]$ is dependent on altogether three conditions. Therefore, the statistics form a four dimensional tensor. In this case, the same disadvantages as in the first approach hold and are even strengthened.

5.7.2.3 Value of prediction

The third approach exploits the relation of the new node value and at the same time restricts the statistics to a two dimensional matrix.

Let Q define the quadrangle consisting of the two direct neighbors v_{i_1}, v_{i_2} and a second degree neighbor v_{i_3} of the node v_j with values A, B, C and N, respectively, with $A, B, C, N \in \mathcal{A}$, see Table 5.3. Moreover, let $D = A + B - C$ be the value that is used for predicting the new value with the two dimensional Lorenzo predictor.

$$\mathbb{P}[N|D = v_{i_1} + v_{i_2} - v_{i_3}] = \frac{\#\{(n_{i_1}, n_{i_2}, n_{i_3}, n_j) \in Q | D = v_{i_1} + v_{i_2} - v_{i_3}, v_j = N\}}{\#\{(n_{i_1}, n_{i_2}, n_{i_3}, n_j) \in Q | D = v_{i_1} + v_{i_2} - v_{i_3}\}}. \quad (5.40)$$

A huge advantage of this approach is the two dimensional statistic that provides, compared to the first two cases, smaller statistics and a faster access. The disadvantage is that we cannot depict many different behaviors with a two dimensional distribution. As the data set is already predicted by the Lorenzo predictor and the node values are residuals, we do not expect a big drawback for the compression rates. If $(A+B-C)$ is small we expect that the prediction works and a transition to a value near to zero is highly probable. In the case of a big $(A+B-C)$, we expect a certain inaccuracy for the quality of the prediction. Moreover, for predicted values the accumulation point is usually zero. In this case the statistics of iMc keep a high amount of information, since the number of transitions that start at a high value are rare. Therefore big values keep more information in the statistic than small values. A further disadvantage is that all possible values of D do usually not coincide with the alphabet $\mathcal{A}$. Therefore, the size of the statistic matrix can be distinctly bigger than in the standard iMc case but we expect a better encoding performance for the predicted values.

5.7.2.4 Minimal number of starting edge values

In the fourth approach, we again only consider the direct neighbors of the node whose value we want to encode. Moreover, we restrict the encoding to a tree based encoding. We choose the tree based on the Lorenzo predictor traversal with the aim, that the statistics keep more information.

Let T define the triangle consisting of the direct neighbors v_{i_1}, v_{i_2} of the node v_j with values A, B and N, respectively, with $A, B, N \in \mathcal{A}$, see Table 5.3. We select the transition v_{i_1} to v_j if

$$\#\{(v_{i_1}, \cdot)\} < \#\{(v_{i_2}, \cdot)\}$$

and v_{i_2} to v_j else. Based on the Lorenzo predictor and its strategy to pass the graph, we try to select a tree that introduces as much information as possible in the statistics.

The first two approaches will be part of future work and will not be investigated here. The main obstacle is that for encoding as well as for decoding the statistics has to be accessed very fast since it is dependent on the current node value, which usually changes at every step. For three and four dimensional statistics, we assume higher access times. Therefore, an application in the field of long time archiving seams meaningful. The application of this strategy can be enabled e.g. by a flag in the compression stage.

5.7.3 Graph differences

In this section we investigate a generalization of the tree differences from Section 5.7.1, where we set a theoretical focus on the compatibility of prediction method and encoding. Basically, graph differences can be seen as a general approach for all types of direct neighbor based prediction methods. These considerations are notably of interest when dependencies in the predicted data are to expect as will be further detailed in Chapter 6. When we model such dependencies, we can exploit them in a statistical encoder. Especially in the case, where we weave the prediction and the encoding, the graph differences can help to find a successful combination. Moreover it represents a mathematical foundation for the entropy calculation after an application of graph based prediction methods. The entropy calculated for the model problem 1 by Formula (3.24) results in a value of $H = 1.4948$. The crucial question is, whether there is a prediction method that reduces the entropy.

Let node values are $a_i, a_j \in \mathcal{A}$, $i, j \in \{1, ..., \#V\}$ and the edge values $u_k \in \tilde{\mathcal{A}}$ for all $k \in \{1, ..., 2 \cdot \#E\}$ and the node i be a direct neighbor of node j. We investigate a random process of the following form which corresponds to a difference method on a graph:

$$a_i = u_k + a_j, \text{ with } i \in N(j),$$

We distinguish two cases. In the first one we assume that are no dependencies left after prediction and therefore, the predicted values are independent. In the second case the residuals are still dependent. The dependencies can be exploited by an entropy encoder. First, we assume that the random variables u_k are on the one hand independent of s_j and on the other hand independent on all other u_l with $k \neq l$.

$$
\begin{aligned}
\mathbb{P}_{X|X}[X_n = s_i | X_{n-1} = s_j] &= \mathbb{P}_{X|X}[X_n = u_k + s_j | X_{n-1} = s_j] \\
&= \mathbb{P}_{Y|X}[Y_n = u_k | X_{n-1} = s_j] \\
&= \mathbb{P}_Y[Y_n = u_k]
\end{aligned}
$$

In Figure 5.23 the values of the edges are listed as the difference of the ending and the starting node. As we focus on undirected graphs there are two values per edge.

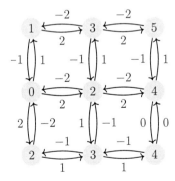

Figure 5.23: Model problem 1 with two values per edge.

The relative frequencies of u_i are listed in Table 5.4.

u	$\#$	$\mathbb{P}[u]$
-2	5	$\frac{5}{24}$
-1	6	$\frac{1}{4}$
0	2	$\frac{1}{12}$
1	6	$\frac{1}{4}$
2	5	$\frac{5}{24}$
Σ	24	1

Table 5.4: *Distribution of the edge values based on the differences between neighbored nodes.*

For the small example the entropy per value is:

$$H = \sum_{u \in \bar{A}} p_U(u) \log_2(p_U(u))$$

$$= 2.2417$$

In the following we assume that even after one up to several prediction steps there are still dependencies that can be described statistically. An entropy can only be calculated meaningfully if mean ergodicity is ensured. Accordingly we have to ensure that the induced Markov chains for the given dataset and all iterates are mean ergodic.

Crucial for an iterative application of graph differences is their construction. Let a connected graph $G = (V, E)$ with at least two nodes be given. There are several possibilities to define differences on a graph. We consider the one specified in the following definition.

Definition 5.38 (Deduced graph). *Let a graph $G = (V, E)$ with the node value vector $v \in \mathcal{A}^n$, $n = \#V$ and an alphabet $\mathcal{A} = \{v_i | i \in \{1, ..., n\}\}$ be given. We match to every edge the absolute difference of the starting and the ending node:*

$$u = |v_{end} - v_{start}| = |v_{start} - v_{end}| \in \mathcal{A}^{(1)},$$

with a new alphabet $\mathcal{A}^{(1)}$. We construct a graph $G^{(1)} = (V(G^{(1)}), E(G^{(1)}))$ with every node $n_i^{(1)} \in V(G^{(1)})$ being dedicated to an edge $e_i \in E$ and the node value vector consists of the edge values of the graph G. The new connectivities will be constructed on the basis of the connectivities of G as follows. If an edge starts or ends in a node it is connected to all other edges which start or end in this node. If two edges $e_i, e_j \in E$ are connected in G, the corresponding nodes $n_i^{(1)}, n_j^{(1)}$ are connected in $G^{(1)}$. We call $G^{(1)}$ the deduced graph of G.

For Model Problem 1 we get the graph like in Figure 5.24.

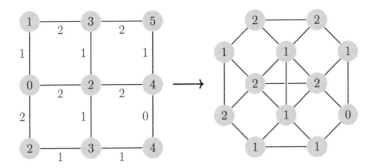

Figure 5.24: *Mesh with edge weights being the absolute differences and its deduced graph.*

Remark 5.39. *The deduced graph is also called line graph in the context of graph theory [32].*

Definition 5.40. *A distribution is called symmetric at $\mu \in \mathcal{A}$ if*

$$\mathbb{P}[X = s_i + \mu] = \mathbb{P}[X = -s_i + \mu]$$

with $-s_i, s_i \in \mathcal{A}$ for all $s_i \in \mathcal{A}$.

The distribution of the edge values is symmetric for every edge $\langle n_k, n_l \rangle \in E$, $n_k, n_l \in V$ with value $s_i = v_l - v_k \in \tilde{\mathcal{A}}$ the edge $\langle n_l, n_k \rangle$ has the value $-s_i = v_k - v_l \in \tilde{\mathcal{A}}$, the alphabet $\tilde{\mathcal{A}}$ is chosen as $\{s_i | i = 1, ..., 2\#E\}$, and the probabilities are determined by relative frequencies. For a symmetric distribution it does not matter if we save the sign of each nonzero value together with the value or in a separate array as proven in the following lemma.

Lemma 5.41. *The entropy of an independent random process X with a finite number of states $s_0, ..., s_n \in \tilde{\mathcal{A}}$ and a symmetric distribution at 0 can be expressed by Equation (5.41) where $H^{(1)}$ is the entropy of the random process X.*

The proof of this Lemma can be found in the appendix.
By omitting the sign of the edge value we ignore the direction of the edge. This information can be expressed by not more than one bit per edge with nonzero value and has to be added to the entropy later. We can split the information about the algebraic sign of a number when the source has a symmetric distribution such as in the following.

$$H = H^{(1)} + \frac{\#\text{Nonzero Entries}}{\#\text{Entries}} \tag{5.41}$$

The deduced graph is irreducible, too, and we can iteratively apply the graph differences. With Lemma 5.41 it is valid to focus on an undirected graph with nonnegative edge values. The construction of the graph in the next step results in a graph with more edges as the number of new edges is

$$\#(E(G^{(1)})) = \sum_{\langle n_i, n_j \rangle \in E} (\deg(n_i) - 1 + \deg(n_j) - 1).$$

In addition, the number of neighbors in two dimensional graphs (shell elements) is as a rule at least two or higher. Therefore, one edge results in at least one edge in the new graph. This behavior is obvious for the Model Problem 1, see Figure 5.24. Such behavior does not increase the amount of information to encode as we do not have to visit all nodes of the deduced graph rather than $n - 1$ nodes where n is the number of nodes of G. For the prediction it is sufficient to use a spanning tree T of the original graph G with only those nodes $n_i^{(1)} \in V(G^{(1)})$ which correspond to edges of T. The induced Markov chain of the deduced graph $G^{(1)}$ is shown in Figure 5.25.

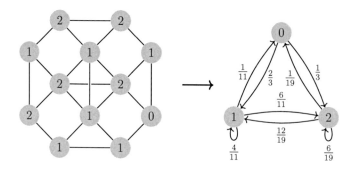

Figure 5.25: Mesh and its induced Markov chain.

For a well-defined investigation of the entropy behavior it is important to verify that the Markov chain of a deduced graph $G^{(1)}$ is also mean ergodic if the induced Markov chain of the original graph G is mean ergodic. In most cases we can ensure the mean ergodicity as well as the ergodicity for $G^{(1)}$, e.g. if the ergodicity of G can be proven via the topology we can deduce the ergodicity for $G^{(1)}$, too.

Lemma 5.42. *The Markov chain that is induced by $G^{(1)}$ is irreducible if G is connected.*

PROOF: With Lemma 5.20 it is sufficient to show that any two nodes of $G^{(1)}$ are connected. Let $n_i^{(1)}, n_j^{(1)} \in V(G^{(1)})$ be given. These nodes corresponds to the edges $e_i \in E$ and $e_j \in E$, respectively. As the original graph G is connected there exists a walk $\langle n_{l_1}, n_{l_2}, ..., n_{l_{k-1}}, n_{l_k} \rangle$ with $\langle n_{l_0}, n_{l_1} \rangle = e_i$ and $\langle n_{l_k}, n_{l_{k+1}} \rangle = e_j$. Define $n_{l_0} = n_{l_0}$ and $n_{l_{k+1}} = n_{l_{k+1}}$. The walk

$$\langle n_{l_0}, n_{l_1}, ..., n_{l_k}, n_{l_{k+1}} \rangle$$

exists on the graph G traversing edges: $e_i, e_{i_1}, ..., e_{i_{k-1}}, e_j$. So there is a walk

$$\langle n_i^{(1)}, n_{i_1}^{(1)}, ..., n_{i_{k-1}}^{(1)}, n_j^{(1)} \rangle$$

in the graph $G^{(1)}$. Thus the graph is connected and the induced Markov chain is irreducible. □

The next step is to show the aperiodicity of an induced Markov chain to the graph $G^{(1)}$. Unfortunately this does not apply to every underlying graph G which induces an ergodic

Markov chain as we can see in Figures 5.26 and 5.27, although we can with Lemma 5.42 ensure mean ergodicity in the context of connected graphs.

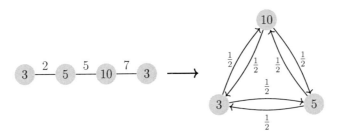

Figure 5.26: Graph G and its induced Markov chain.

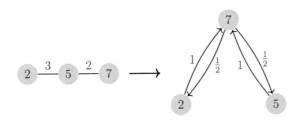

Figure 5.27: Deduced graph $G^{(1)}$ and its induced periodic Markov chain.

In the next step we show that circles with odd length in G will remain in $G^{(1)}$. Together with Theorem 5.24 it follows that the induced Markov chain of $G^{(1)}$ is also aperiodic.

Lemma 5.43. *Let G be a connected graph and $G^{(1)}$ the graph constructed like in Definition 5.38. If G has a circle with odd length, $G^{(1)}$ will also have a circle of the same odd length.*

PROOF: Let $\langle n_{i_1}, n_{i_2}, ..., n_{i_l} n_{i_1} \rangle$ be an odd circle of length l in G with $n_{i_j} \in V$ for all $j = 1, ..., l$. In edge notation it is $\langle e_{j_1}, e_{j_2}, ..., e_{j_{l-1}}, e_{j_l} \rangle$ with $e_{j_s} = \langle n_{i_s}, n_{i_{s+1}} \rangle$ for all $s = 1, ..., l-1$ and $e_{j_l} = \langle n_{i_l}, n_{i_1} \rangle$. Thus the walk $\langle \tilde{n}_{j_1}, \tilde{n}_{j_2}, ..., \tilde{n}_{j_l}, \tilde{n}_{j_1} \rangle$ is part of $G^{(1)}$ with $\tilde{n}_s \in V(G^{(1)})$ and has length l. □

Proposition 5.44. *If the undirected and connected graph G has a circle with odd length the induced Markov chain is ergodic and $G^{(1)}$ also induces an ergodic Markov chain.*

Lemma 5.45. *Let G be a connected undirected graph with at least one node $n_i \in V$ with $deg(n_i) \geq 3$, then the induced Markov chain of the graph $G^{(1)}$ is aperiodic.*

PROOF: By definition of the graph $G^{(1)}$ there are at least three edges $e_{i_1}, e_{i_2}, e_{i_3} \in E$ which start or end in the node $n_i \in V$ with $deg(n_i) \geq 3$. These three edges will become a triangle in $G^{(1)}$ which is a special case for a circle with odd length. □

Hence, only for graphs which consist of nodes with a degree smaller than three it is not possible to construct a chain of deduced graphs that are all mean ergodic. There are two possibilities: Either the graph consist of circles with even length or it consists of linear chains with no intersections.

Remark 5.46. *When we consider the time dependence we get additional connectivities between time steps. Therefore, we can increase the number of edges for each node in each graph and, thus, obtain a node with a degree of at least three for all connected graphs with at least three nodes.*

For further theoretical investigations, we only consider graphs for which we can guarantee the ergodicity of an induced Markov chain by their topological properties. For these graphs the construction of the line graph can be applied on $G^{(1)}$ while we can ensure that G, $G^{(1)}$ and $G^{(2)} = (G^{(1)})^{(1)}$ induce an ergodic Markov chain. So we can build a chain of graphs inducing ergodic Markov chains.
For these chains we can calculate the entropy and determine a minimum that will give us the best possible compression ratio for this purpose.

Remark 5.47. *A possible application of graph differences is the following scenario. Instead of determining a minimum spanning tree based on the edge weights that are the absolute differences of the corresponding node values, see [107], we select the edges in a way that the tree generates the minimum entropy in the sense of iMc encoding.*

The strategy described in Remark 5.47 creates a direct connection between prediction and encoding.

Remark 5.48. *Our strategy to use graphs for the encoding enables to investigate the optimal combination of a prediction-encoder in several ways. We achieve better compression rates if we do not examine the predictor and the encoder as two separate components but use the information mutually.*

This assumption is to be underpinned by the following observations. In Figure 5.28, the optimal tree in the sense of iMc is displayed with one tree difference. It is obvious, that

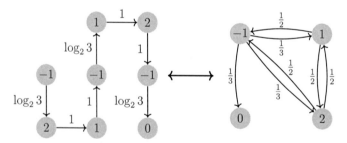

Figure 5.28: *Costs for encoding the graph based values with a nonadaptive iMc encoder for the tree of Figure 5.3. The costs are determined by the negative logarithm to base 2 of the transition probabilities. A transition with probability one is free.*

this tree is far from optimal for a combination of one tree difference and iMc encoding.

The costs for saving the transitions for the third till ninth value would be $9 = \lceil 8.75 \rceil$ bits. Additionally, the root node value for an application of the tree differences as well as the first predicted value has to be stored as side information. The optimal tree for a combination of the application of one tree difference and iMc encoding can be determined based on the deduced graph $G^{(1)}$, see Figure 5.29. The induced Markov chain together

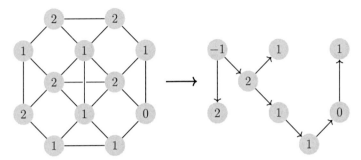

Figure 5.29: *The deduced graph is the basis for finding an optimal tree for the combination of tree prediction and iMc encoding. Since we give each edge a direction, we can determine the sign of each node. The tree and the statistics for iMc coding can be found in Figure 5.30.*

with the costs for encoding the edges can be found in Figure 5.30. In the optimized case

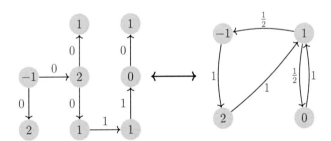

Figure 5.30: *Costs for encoding the graph based values with a nonadaptive iMc encoder for the optimal tree of Model problem 1 with one tree difference. The costs are determined by the negative logarithm to base 2 of the transition probability. A transition with probability one is free.*

the costs for saving the transitions for the third till ninth value would be 2 bits. In the case of adaptive iMc encoding, the costs would be even 1 bit.

Bottom line, it makes sense to investigate the combination of the prediction and the encoding phase. Considering the law of large number we expect that for big meshes the choice of a tree is less important compared to examples of small graphs that are investigated in this section.

5.8   Differentiation of induced Markov chains to other approaches

In this section, we delimit the iMc to Prediction by Partial Matching (PPM), Dynamic Markov Coding (DMC), Bayesian Networks, Hidden Markov Models (HMM), and Markov Random Fields (MRF). These state-of-the-art approaches have already been discussed in Section 3.3.3.

5.8.1   Prediction by Partial Matching and Dynamic Markov Coding

PPM and DMC are two methods that use a Markov model to encode a one dimensional stream, see Section 3.3.3.3 and 3.3.3.5. The modeling is restricted by the consideration of dependencies in time direction. For graph based data, of course, we could perform a mapping into a one dimensional stream and assume that the value at position $i + 1$ is dependent on the value at position i. The properties of the mapping will strongly influence the success of a PPM or a DMC approach. A central aim of the induced Markov chains is to exploit the topological relations which is not possible for Dynamic Markov Coding and Prediction by Partial Matching and which is independent of a mapping into a one dimensional stream.

5.8.2   Bayesian Networks

In Section 3.3.3.7, the special cases of graphical models, namely Bayesian Networks and Hidden Markov Model are stated.

As a machine learning approach Bayesian Networks need a certain training set for initialization. Since the statistics are usually not stored in the Bayesian Networks case, in contrast to the iMc case, a direct decompression of a single time step is not possible. One big difference between the application of iMc and general Bayesian networks is, that we usually only consider one instance of a graph. Therefore, the number of learning set would be maximal the number of time steps minus one. Usually, the size of the training set is not sufficient. Bayesian networks are constructed in a way that if we already know a node value, although one of its ancestors is still unknown, we can benefit from the belief propagation [45, 67]. Since we have a certain order to traverse the graph, such situations are not possible in the case of iMc. When we apply our iMc in combination with the Lorenzo predictor, we face directed acyclic graphs (DAG), which are common in Bayesian Networks. An example for such a DAG can be found in Figure 3.12b, which corresponds to the combination approach 1 and 4 of Section 5.7.2, and Figure 3.12b, which corresponds to the combination approach 2 and 3 of Section 5.7.2.

The iMc has not the drawback of Bayesian networks of no random access and updating the statistics in the decoding stage, see Remark 3.73. Therefore, iMc encoding is meaningfully applicable if we have only one or few observations that are usually equal to the number of time steps. Especially, when compressing sets of simulation results, a certain machine learning approach makes sense and is very interesting. However random access is an important property and has to be guaranteed for the usual workflow.

The Hidden Markov Model is a special case of Bayesian networks. The main difference to our iMc approach is, that we assume, that the observed variables themselves have the Markov property. Therefore, we do not need an extra layer of hidden variables. Moreover, the Hidden Markov Model is usually used in machine learning and has the same drawbacks as the Bayesian Networks for data compression applications [81].

5.8.3   Markov Random Fields

The Markov Random Fields as well as the Bayesian Networks are part of the graphical models.
A major difference to iMc coding is that the model assumes a predefined distribution model, such as the Ising model. Therefore, it can be adapted to a certain use case, but cannot be used as a black box method. This is different for the iMc. There is no assumption that a certain distribution of the data set takes place in advance. Moreover, iMc is applicable for well as well as for badly predicted data sets. Due to the time complexity of MRF, an application only makes sense if the alphabet is very small, see Remark 3.77. Again, the iMc can also handle this situation, since the transition probability matrix is stored in a sparse matrix format and therefore only a number of edges $\#E$ can have nonzero entries that are independent of the alphabet size.
A further difference is that the Markov Random Field is defined in an undirected graph. A graph or grid from a finite element simulation also has no directed edges. In this understanding, it would be a meaningful choice. But we decided to use directed edges, because we have a certain direction of encoding and we do not want to pass the tree several times for decoding, while the decoding run time is a crucial parameter.

Conclusion:

In this chapter, we introduced the induced Markov chain (iMc) encoder, see Section 5.1, that is the first encoder based on a graphical model that can meaningfully be applied to big alphabets. Especially the time complexity examined in Section 5.2 shows that an application is possible. The iMc encoder is an asymmetric two-pass encoder that uses an arithmetic encoder to encode with a distribution based on the Markov induced chains presented in Section 5.4. The idea of the induced Markov chains is to translate a topological relation into a value based relation. In addition, we introduce an adaptive version that exploits the drawback of storing the statistics by eliminating already processed elements from the statistics. To our knowledge, it is the first application that updates the statistics by eliminating elements already processed. We adapt our distribution to fit the future and not to the past, see Remark 5.11.
In Section 5.5, we prove a bound on the Kullback Leibler divergence of the stationary distribution of the iMc and the empirical distribution, which are the relative frequencies, solely based on the topology of the underlying graph. We can interpret the divergence of the empirical and the stationary distribution as follows:
We perform a random walk on a given graph and set the next value based on the transition probabilities of the iMc. The stationary distribution is the mean distribution of the

Markov process. Therefore, we can investigate how close the empirical distribution, which can be seen as the distribution of a snapshot of the random walk, fits to the stationary distribution of the random walk. The closer the distributions are, the better the iMc model fits to the data set.

The Kullback Leibler Divergence gives us the additional number of bits needed to compress a data set with an optimal code for one distribution, but actually was generated based on a different distribution. Since the bound on the Kullback Leibler Divergence, see Theorem 5.36 is very tight, the iMc model fits very good, especially when the graph has a certain regularity.

In Section 5.6, we motivate the usage of a breadth first search technique to determine the tree for the iMc encoder. We also show that this is not necessarily the best choice. In Section 5.7, we combine the iMc encoding with a tree prediction and Lorenzo prediction. In the latter case, the iMc model is a Bayesian network, since it also uses dependencies between non-adjacent nodes.

We close this chapter with a distinction of the induced Markov chain to graphical models on the one hand and encoders that interpret all data as a one dimensional stream on the other.

Chapter 6

Evaluation

In this section, we evaluate our PPCA method in combination with the iMc encoder and compare it with state-of-the-art methods, see Section 3.3.2. Two important aspects of our evaluation are compression rates and runtimes.

We start this chapter with the methodology of our evaluation and then introduce the benchmark data sets. We then focus on the crucial parameters that are freely selectable from PPCA methods. After specifying the parameters, we examine the PPCA methods in combination with Rice, zlib and iMc encoder based on compression factor and run time. We close this chapter with an examination of the PPCA Online method as a core technology in a SDMS.

6.1 Methodology

We evaluate our strategies for the compression of sets of simulation results using three models. All three models were simulated for robustness analysis with parameter variations, and show only minor changes in the initial state. Such data sets are generated, for example, for robustness analysis using DIFFCRASH [123].

We investigate whether the PPCA methods belong to a class of methods that can benefit from the existence of several similar simulation results. For this purpose we determine a learning factor which relates the compression rate $CF(1)$ of the PPCA Offline method based on a model to the compression rate $CF(n)$ of the respective PPCA method based on all n models as follows:

$$\mathcal{L} = \frac{CF(n)}{CF(1)} \tag{6.1}$$

If $\mathcal{L}$ is greater than 1, regardless of the simulation result on which the compression rate $CF(1)$ was determined, we say that the method can learn from the existence of an array of simulation results. In the PPCA Online case, we determine the learning factor by comparing the compression rate of the PPCA Online method on all data sets with the compression rate of the PPCA Offline method on the first data set.

As part of the PPCA procedure's critical parameters and heuristics were introduced (see Section 4.3.4), the effect of which on the performance must be determined and the respective choice must be justified. Furthermore, the compression rates for the coordinates and other simulation results defined on the nodes or elements of the Finite Element Grid are

Table 6.1: *Excerpt of precisions used for evaluation of Chrysler Silverado, OEM 1 (PAM), OEM 2 (LS-DYNA). The complete list of precisions can be found in Tables C.1, C.2, and C.3 in the appendix.*

Variable name	Defined on	Precision		
		Silverado	OEM1	OEM 2
Coordinates	Nodes	0.1	0.1	0.1
Thickness (THIC)	2D Shells	0.01	0.01	0.01
Effect. Plastic Strain EPLE	3D Elements	0.001	0.01	0.01

determined and compared with the current state of the art. The state of the art is represented by the current variants of FEMZIP-L 10.61 and FEMZIP-P 10.61, each of which represents specialized variants for lossy compression of simulation results in LS-DYNA d3plot and Pam-Crash DSY respectively. For the compression using FEMZIP, as well as our methods, a parameter file was generated for each simulation result, in which the quantums, see Section 3.3.1, for the lossy compression of nodes and element variables and the coordinates were specified, see Table C.1, C.2, and C.3. The maximum error caused by the lossy compression can be estimated according to Equation (3.28). For FEMZIP-P as well as FEMZIP-L, the "-L4" option is recommended to be used by SIDACT, since it provides good compression factors as well as fast postprocessing run times.

The freely available software 7zip was applied to the lossy compressed and decompressed model. By applying 7zip to the reconstructed model, the positive effect of lossy compression on the compression rate is achieved. The main difference between 7zip and FEMZIP is therefore the understanding of the data format and the resulting special adaptation of the compression methods.

Two application scenarios are tested for our PPCA procedure. In the first case, all simulation results are already available and should be compressed. In this case we use the PPCA procedure in Offline mode. This case can occur during long-term archiving, where particularly high compression rates are desired. The second case describes a scenario in which one simulation result after the other becomes available. This use case occurs, for example, if we start and organize simulations with a SDMS. With SDMS, we compress each new data set with the previous knowledge of the previously compressed results. Thus, we apply the PPCA procedure in Online mode except for the compression of the initial simulation result. We combine both the PPCA Offline and the PPCA Online method with the encoders zlib with level 9 [111], Rice, see Section 3.3.3.4 and [145] and our graph-based iMc encoder, see Section 5.1 as residual encoder. The iMc encoder is only applied to variables assigned to nodes, e. g. coordinates and velocities, as part of our evaluation. For element variables the Rice encoder is used in the iMc case.

In our evaluation, we focus on the size of the compressed file since we need to save the geometry per part only once for the entire data set. If only the compressed results for the coordinates or other time-dependent variables were compared instead, this advantage of the PPCA method would be ignored. Moreover, focusing on the compressed file sizes allows a comparison with the compression tool 7z.

Table 6.2: Properties of the three benchmark data sets.

Model	Silverado	OEM 1	OEM 2
Format	d3plot	DSY	d3plot
Number of Parts	682	5,356	6,625
Number of time steps	152	141	62-77
Number of nodes	942,749	1,785,996	5,202,639
Number of 1D elements	2,697	318,257	98,017
Number of 2D elements	873,191	1,547,032	4,623,446
Number of 3D elements	53,293	105,255	1,219,252
Simulation span	150ms	140ms	120-150ms

6.2 Benchmark data sets

In the following section we briefly describe the data sets on which we have applied our compression tools. An overview of the properties of the benchmark data sets can be found in Table 6.2.

6.2.1 Chrysler Silverado

The LS-DYNA™ [84] model of the Chrysler Silverado models a frontal crash and is provided free of charge by the National Highway Traffic Safety Administration (NHTSA), see [98]. The model consists of 929181 elements, 682 parts divided into 34 1D elements, 567 2D elements and 81 3D elements. The model was simulated 50 times with sheet thickness variations without geometric changes. 152 time steps with a sampling rate of 1ms were written out. Thus, the temporal resolution is twice as high in comparison to simulation results, which are generated nowadays by automobile manufacturers. On the other hand, the number of elements is smaller by a factor of 5. The ranges for the x-, y-, and z-coordinates are $[-5,929; 296]$, $[-1,234; 1,234]$, and $[-110; 1,922]$ millimeter (mm), respectively. Although the models have an identical initial geometry, the simulation results already differ considerably from half the time steps, see Figure 6.1. This is due to the non-robust behavior of the simulation model in combination with a strong variation in sheet thicknesses of up to 20% for crash relevant components. If we perform a decomposition of the Silverado model into parts and assign nodes that occur in multiple parts to each part in which the node occurs once, we generate an overhead of 4.2%.

In the further course of this evaluation we take a closer look at two parts

1. driver's door, see Figure 6.3 and

2. hood, see Figure 6.2.

The hood is part of the crash area that is drastically deformed. In contrast, rotation and translation work mainly on the driver's door. Hence, we have two different use cases for our compression strategies. Both parts are large in number of nodes and elements, see Table 6.3. Therefore, it is important that our compression performs well on those examples to obtain a good compression factor for the entire Chrysler Silverado model.

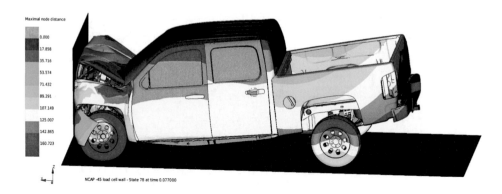

Figure 6.1:   The maximal node distance in mm between the first simulation result and all other 49 results for the Chrysler Silverado data set.

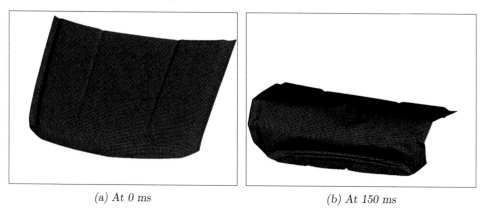

(a) At 0 ms   (b) At 150 ms

Figure 6.2:   The hood of Chrysler Silverado data set for a fixed viewpoint.

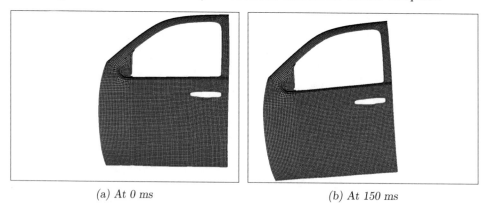

(a) At 0 ms   (b) At 150 ms

Figure 6.3:   The driver's door of Chrysler Silverado data set for a fixed viewpoint.

Table 6.3: *Properties of the hood and the driver's door from the Chrysler Silverado benchmark data set.*

Part name	Part id	Number of nodes	Element type	Number of elements
Hood	2,000,349	13,802	2D Shell	13,604
Driver's door	2,000,457	8,375	2D Shell	8,069

Table 6.4: *Components of one simulation result from OEM 1 and its uncompressed size in MB.*

State information	Uncompressed size
Coordinates	2,882
Shell variables	1,664
Solid variables	57
Sum	4,603

Initial geometry	Uncompressed size
Coordinates	20
Connectivities	40
Sum	61

6.2.2 OEM 1

The OEM data set 1 models an ODB (Small overlap) crash [41] with a scatter on the angle at which the car hits the barrier. These models were simulated with ESI Pam-Crash™[40] and written out in DSY format. The data sets have no geometric variation, but before each run the automated process of simulation re-connects the components responsible for the stability of the vehicle in the simulation. Such connections are usually modeled as 1D parts. The simulation result contains 141 time steps, 1,785,996 nodes, 318,257 1D elements, 1,547,032 2D elements and 105,255 3D elements. The sampling rate is 1ms. The model is divided into 5356 parts, 3197 1D parts, 2028 2D parts and 131 3D parts. The sizes of the particular data sets of the investigated array of simulations are listed in Table 6.4. Further, 90 MB are used for header informations, look-up tables and global variables. The ranges for the x-, y-, and z-coordinates are $[-6,959; 4,852]$, $[-7,438; 2,986]$, and $[-6,950; 4,767]$ millimeter (mm), respectively. We have 24 simulation results available for this dataset The overhead due to the duplication of nodes that occur in multiple components is 18.1%.

6.2.3 OEM 2

The OEM data set 2 also models an ODB (Small overlap) crash [41]. For this example, the sheet thicknesses in the crash relevant area, i.e. the engine compartment, were varied. The model was simulated with LS-DYNA™[84] and written in d3plot format without any geometric variation. The simulation result consists of 5,202,639 nodes, 98,017 1D elements, 4,623,446 2D elements and 1,219,252 3D elements. All models have either 62 time steps or 77 time steps. The sampling rate is 2ms. The LS-DYNA model is divided into 838 1D parts, 5,306 2D parts and 481 3D parts. This makes it by far the largest model in our test series. The number of time steps are in the order of magnitude in which crash test simulations are currently being tendered by OEMs. The ranges for the x-, y-, and z-coordinates are $[-2.4804E+05; 1.8733E+05]$, $[-1.4911E+05; 1.3248E+05]$, and $[-3.7158E+04; 1.5786E+05]$ millimeter (mm), respectively. For the OEM 2 data set, we have 10 simulation results present for this dataset. The overhead generated by the duplication of nodes occurring in multiple components is 22.7%.

6.3   PPCA's crucial parameters

The effectiveness of the PPCA procedures from Sections 4.3.2 and 4.3.3 is influenced
by several adjustable parameters. These crucial parameters have already been examined
theoretically in Section 4.3.4 and will empirically be examined in this section. Since
the choice of one parameter can influence the optimal choice of another parameter, e.g.
internal precision and the number of amendment steps, we consider them together in
our study if necessary. Furthermore, the process for determining the number of principal
components will be investigated. We conclude the section with an examination on a lower
bound for the number of nodes. All parts having less nodes than the given bound will
not be compressed using the PPCA methods.

6.3.1   Number of principal components

The number of principal components of the PPCA methods is determined by an optimization process, see Section 4.3.4.1. Assumptions are made about the relationship between
the compression rate and the number of principal components used, so that a considerable
number of evaluation operations can be saved. Whether these assumptions are correct is
to be checked exemplary using the simulation result set with 3 Chrysler Silverados.

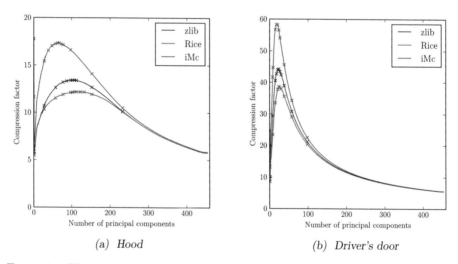

(a) Hood   (b) Driver's door

Figure 6.4: *Plot of compression factor in respect to the number of principal components
for the x coordinates of the driver's door and the hood from 3 Chrysler Silverados [98]
applying the PPCA Offline method. The curves show the compression factor evaluated
for more than 450 different number of principal components. The 'x' markers are points
that are evaluated as part of our PPCA optimization process.*

Figure 6.4b shows the compression rate plotted for the driver's door as a function of the number of principal components used. As described in Section 4.3.4.1, time differences, see [94], are applied in the case of zero principal components. Especially in cases where the simulation result has a high temporal resolution, as with the Silverado data sets, this method already achieves good compression rates. For the optimization process, the case of zero principal components is not considered at first. Once the optimum has been found using the Fibonacci Search method, it is compared with the special case of zero principal components and the approach leading to better compression factors is selected.

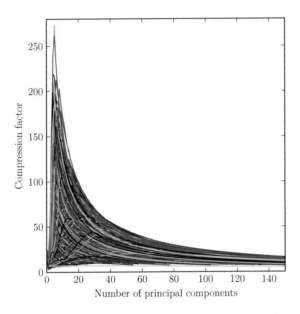

Figure 6.5: Number of principal components and the corresponding compression factor for iMc encoder and the x-coordinated for all parts with at least 40 nodes, see Section 4.3.4.3.

Figure 6.5 shows the compression rate curves as a function of the number of principal components for the x-coordinates of all parts. The curves are either unimodal or have a global curve that justifies the optimization approach. In our optimization approach, the area of few principal components is scanned finer than the area of many principal components. Curve shapes justify this approach because for all curves the area where many principal components are used for prediction is flat, but there are several curves that are very steep in the area of a few principal components.

Table 6.5 compares the compression rates for a full optimization, for which any number of principal components have been tested with those of our optimization approach. The run time advantage of our quasi-optimization justifies its use, since instead of hundreds of evaluations, only 11-13 compression rate evaluations have to be performed. Table 6.5 lists the compressed sizes for iMc, Rice, and zlib encoder together with the deviations

Table 6.5: *Achieved compression sizes of a PPCA Offline run for 3 Chrysler Silvera-dos. The full optimization evaluates every number of principal components in the range* $[0, ..., 453]$ *in contrast to the PPCA optimization, see Section 4.3.4.1, that evaluates 10 - 15 different evaluations. Here, an internal precision of* 2^{-8} *was applied.*

Encoder	Full optimization	PPCA optimization	Deviation
iMc	174,921,342	175,186,499	0.15%
Rice	194,930,730	195,153,124	0.11%
zlib	219,423,853	219,802,977	0.17%

Table 6.6: *Achieved compression factors on the x coordinates of the driver's door for our PPCA optimization, see Section 4.3.4.1, and an optimization which evaluates every possible number of principal components for an application of PPCA Offline on 3 Chrysler Silverados and internal Precision* 2^{-8}. *The number in braces are the respective number of principal components. The uncompressed size is 5,092,000 Bytes.*

Encoder	Full optimization	PPCA optimization	Deviation
iMc	58.36 (18)	58.36 (18)	0%
Rice	38.81 (26)	38.64 (24)	0.44%
zlib	44.40 (23)	44.17 (24)	0.52%

between our PPCA optimization approach and the full optimization. The deviations for all applied encoders are less than 0.2% and therefore negligible. For the driver's door and hood examples, our optimization algorithm leads to deviations in the compression rate compared to a full optimization of less than 0.2% and less than 0.6%, respectively. The compression rate drawback is negligible in comparison to the savings in run time, since we evaluate between 11 and 13 different numbers of principal components in contrast to over 450.

6.3.2   Internal Precision and amendment steps

The internal precision is used to convert floating-point numbers to integer numbers by quantization, see section 3.3.1. The internal precision is applied to the mean values, coefficients and principal components and determines, among other things, how good a prediction can be with the PPCA method. We consider the internal precision in combination with the number of amendment steps.

On the one hand, there are internal precision requirements to guarantee the convergence of the amendment steps, see Theorem 4.14, on the other hand, the principal components for inaccurate internal precision are usually further away from being orthogonal after lossy compression and decompression than in the case of high accuracy. Thus, it is to be expected that in such a case a higher number of amendment steps would be useful. The amendment steps are only used in the PPCA Online procedure. Our prototypical use case is three Chrysler Silverados, which are added one after the other. This means that we have one application of the PPCA Offline method followed by two applications of the PPCA Online method.

Table 6.7: *Achieved compression factors on the x coordinate of the hood for our PPCA optimization, see Section 4.3.4.1, and an optimization which evaluates every possible number of principal components for an application of PPCA Offline on 3 Chrysler Silverados and internal Precision 2^{-8}. The number in braces are the respective number of principal components. The uncompressed size is 8,391,616 Bytes.*

Encoder	Full optimization	PPCA optimization	Deviation
iMc	17.77 (0)	17.77 (0)	0%
Rice	12.20 (113)	12.18 (120)	0.16%
zlib	13.43 (101)	13.41 (108)	0.15%

The size of the compressed file was determined depending on the precision and the number of amendment steps. This approach was applied together with the zlib, iMc and, Rice encoders and the results are shown in Figures 6.6, 6.7, and 6.8. An evaluation was made for all points of the rectangular grid shown in the figures. The curve is similar for all three encoders. The minimum is reached for iMc and zlib with two amendment steps and the use of an internal precision of 2^{-8}. For the Rice encoder 3 amendment steps are optimal. For all cases it is obvious that the first amendment step already has the greatest influence on the overall size and shows the importance of the amendment steps within the PPCA Online procedure. In addition, the curves justify our assumption that amendment steps are more important in the case of inaccurate internal precision than in the case of high internal precision. The overall sizes and thus the compression rates differ only insignificantly for the application of more than two amendment steps. On this basis, we have defined a default value of 3 amendment steps as the default setting for all further tests of the PPCA Online procedure. When using 3 amendment steps, the internal precision of 2^{-8} is optimal for the Chrysler Silverado data set. The difference to the results for the precision 2^{-9} is small at least for the Rice encoder, see Table 6.8. In addition, for the PPCA Offline method applied to 3 Silverado simulation results, it turns out that the Rice encoder provides better results for the precision 2^{-9} than for the precision 2^{-8}.

Table 6.8: *Compressed sizes for three Chrysler Silverados achieved by PPCA methods for internal precision of 2^{-8} and 2^{-9}.*

PPCA mode	precision	iMc	Rice	zlib
Offline	2^{-8}	175,186,499	195,153,124	219,802,977
	2^{-9}	176,406,918	192,715,795	219,947,185
Online	2^{-8}	191,724,031	207,730,523	235,152,499
	2^{-9}	197,401,166	212,047,428	241,895,003

Therefore, we have additionally considered the OEM 1 example to determine the default value on the internal precision for all further evaluations. Here the situation is different from that for the Chrysler Silverado models, see Table 6.8. For PPCA Online as well as PPCA Offline and all encoders, the compressed sizes are smaller with an internal precision of 2^{-9} in contrast to 2^{-8} for the Silverado data set. For the internal precision of 2^{-9}, the relationship between the size of the database and the simulation specific data sets changes in favor of the database, see Figure 6.9. This is desirable because this information

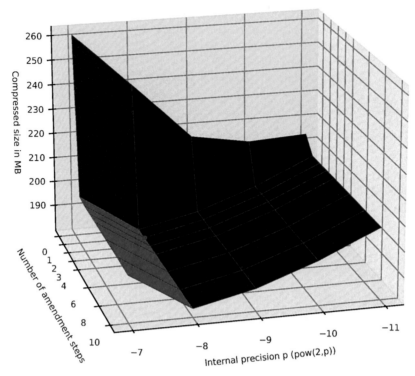

Figure 6.6: *Variation of the number of amendment steps and the internal precision and the respective compressed size applying the iMc encoder. The red dot marks the minimum for internal precision* -8*, i. e.* 2^{-8}*, and 2 amendment steps.*

is available for further application of the PPCA Online procedure. Based on these results, we have decided to use an internal precision of 2^{-9} in the further course of this work.

6.3.3 Small parts threshold

There is a minimum number of 40 nodes for a part from which the application of PPCA procedures is considered. This barrier was introduced in order to reduce complexity, especially for many small parts. This makes sense, because for very small components an application of the PPCA method rarely leads to a better compression rate than, for example, the time difference method, see [94]. By specifying this lower limit, the optimization process for small components can be saved. The influence on the compression rate to exclude such components is small. For the Silverado data set with one and with three simulation results, the total size of the compressed models for a lower barrier of 10, 20, 30 or 40 nodes was determined and summarized in Table 6.10. The difference is only in the range of one thousandth of the compressed size. Therefore, this parameter should not be considered critical unless it is too large and the choice of 40 nodes makes sense.

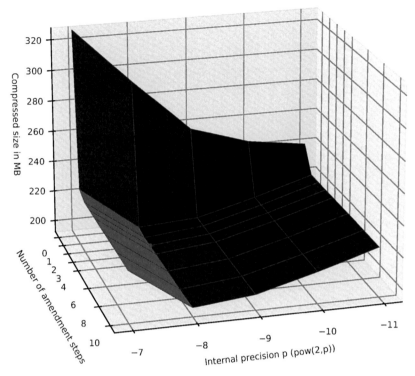

Figure 6.7:   Variation of the number of amendment steps and the internal precision and the respective compressed size applying the Rice encoder. The red dot marks the minimum for internal precision -8, i. e. 2^{-8}, and 3 amendment steps.

6.4   Compression factors

This section examines two points in more detail. First, we empirically test how the compression rates of the PPCA methods are compared to those of the state-of-the-art tool FEMZIP and the universal encoder 7z using the benchmark data sets defined in Section 6.2. Second, we investigate whether PPCA methods are among those that can benefit from the existence of several similar simulation results. We list the compression rates obtained by using the PPCA Online and PPCA Offline methods in combination with the application of the encoders Rice, zlib and iMc to the residual of the PPCA prediction, see Equation (4.33) for the benchmark data sets from Section 6.2. Since the PPCA Online method cannot be used for a single simulation result, there is no value in the result tables.

We compare the results with those of the state-of-the-art tool FEMZIP and the universal encoder 7z. By activating the verbose mode in FEMZIP, the compression factors for individual variables are determined in addition to the compression factor based on the source and target file.

They are different for the use of the iMc encoder. All of these variants have in common that the iMc encoder is only applied to node-based variables such as coordinates, velocities

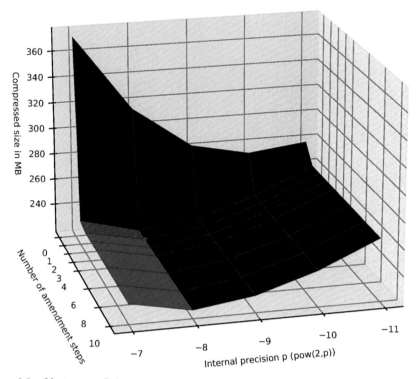

Figure 6.8:   Variation of the number of amendment steps and the internal precision and the respective compressed size applying the zlib encoder. The red dot marks the minimum for internal precision -8, *i. e.* 2^{-8}, *and 2 amendment steps.*

and acceleration. For element based variables the Rice encoder is used in the iMc case. The iMc encoder can either be applied to the node variables of all parts or exclude certain parts, such as parts consisting of 1D elements. In this case, the Rice encoder is also used for the remainder of all parts. We call this combination iMc 2+3D. Furthermore, the iMc encoder can be combined with tree differences, which is the case with iMc T, in case the iMc encoder is also applied to 1D parts, iMc 2+3D T in case the 1D parts are excluded. Thus the application of the iMc encoder is available in four versions. For all results from the previous Section 6.3 the iMc encoder was applied to 1D, 2D, and 3D parts but without tree differences.

For all three data sets, the compression is applied to both a simulation result and the entire set of simulation results. For the application to all simulation results, all results are compressed at once using the PPCA Offline method. To evaluate the PPCA Online method, one simulation result is compressed after the other, so that for a component contained in m simulation results, PPCA Offline is applied once and PPCA Online is applied $m - 1$ times. The results are listed in Tables 6.11, 6.17, and 6.20.

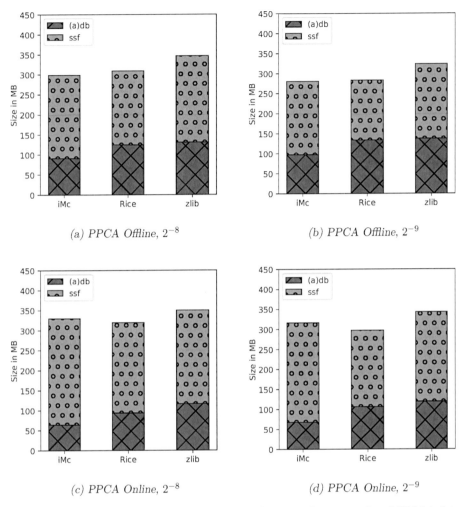

(a) PPCA Offline, 2^{-8} (b) PPCA Offline, 2^{-9}

(c) PPCA Online, 2^{-8} (d) PPCA Online, 2^{-9}

Figure 6.9: *Variation of the internal precision for 3 simulation results of OEM 1 data set applying the PPCA Online methods and PPCA Offline method. The height of the bar is the compressed size as a combination from the data base size in red and the simulation specific files in blue.*

6.4.1 Silverado

An overview of all achieved compression factors for the Chrysler Silverado dataset can be found in Table 6.11. All compression rates are extraordinary high. This is because these simulation results contain time-dependent data whose values were quantized to zero everywhere by a previous application of FEMZIP, see parameter file C.1. The rough quantization was applied to the "velocities", the "accelerations" and the variables defined on 2D el-

Table 6.9: *Compressed sizes for 3 models of OEM 1 data set achieved by PPCA methods for internal precision of 2^{-8} and 2^{-9} in combination with a residual encoder.*

PPCA mode	precision	iMc	Rice	zlib
Offline	2^{-8}	313,290,996	324,651,552	363,793,448
	2^{-9}	294,713,208	297,344,064	339,537,530
Online	2^{-8}	345,537,679	335,355,515	367,410,256
	2^{-9}	332,065,364	311,320,141	359,688,712

Table 6.10: *Achieved compression factor for various lower bounds on the minimal number of part nodes; the PPCA Offline method was applied on 1 and on 3 simulation results of data set Silverado, iMc encoder, and internal precision 2^{-8}.*

Lower Bound	10	20	30	40	50
1 result	140,129,057	140,129,410	140,130,364	140,131,524	140,136,399
3 results	175,058,456	175,065,755	175,122,632	175,186,499	175,293,335

Table 6.11: *Original and compressed file size in bytes applying FEMZIP-L4, 7zip, and our PPCA method in Offline and Online mode on data set Silverado, see Section 6.2.1. The PPCA method is combined with the encoder zlib, Rice, and iMc for the compression of the residual of the PPCA prediction.*

		Size		Compression factor		Learning factor
Number of results		1	50	1	50	50 vs. 1
Original		9,720,868,864	486,043,443,200	1.0	1.0	1.0
FEMZIP-L4		150,960,128	7,606,325,248	64.4	63.9	1.0
7z		821,358,193	40,773,046,856	11.8	11.9	1.0
PPCA Offline	zlib	92,088,799	2,027,098,311	105.6	239.8	2.3
	Rice	81,708,074	1,563,835,540	119.0	310.8	2.6
	iMc	75,742,360	1,503,680,897	128.3	323.2	2.5
	iMc T	77,935,792	1,553,129,653	124.7	312.9	2.5
	iMc 2+3D	75,661,422	1,489,190,144	128.5	326.4	2.6
	iMc 2+3D T	77,903,588	1,527,051,430	124.8	318.3	2.6
PPCA Online	zlib	-	2,519,425,809	-	192.9	1.8
	Rice	-	2,066,006,425	-	235.3	2.0
	iMc	-	2,171,637,780	-	223.8	1.7
	iMc T	-	2,216,051,358	-	219.3	1.8
	iMc 2+3D	-	2,160,414,635	-	225.0	1.8
	iMc 2+3D T	-	2,199,661,152	-	221.0	1.8

ements "element_dependent_variable_1" and "element_dependent_variable_2". Since the compression rates for the variables are very high and have no informative value, we do not investigate them further. The compression rates for the 1D components are also not taken into account, since the 1D variables for all 50 simulations account for less than 82 MB of the total size. Our focus is on studying the compression rate of the complete simulation results, the coordinates and the variables defined on 2D and 3D elements. Due to the efficient compression of the variables, which have been quantized to zeros and make up almost 210 GB of the original size of 453GB, the compression rates are almost twice as high as would be expected. However, all three methods FEMZIP, 7z and the PPCA methods benefit from big zero blocks, so that we assume that the compression factors are a factor 2 too large, but the ratio between the compression strategies is not influenced by the roughly quantized variables.

The free compression tool 7z achieves a compression factor of almost 12 on a simulation result from the Silverado data set. We would expect a compression factor of 6 if the zero quantized data were not available. The compression tool FEMZIP achieves a compression factor 5.4 times higher than 7z with the recommended "-L4" option. The difference is due to FEMZIP's understanding of the format, while 7z interprets the simulation result as a byte stream. FEMZIP, therefore, is able to apply a specialized procedure for each data block. Despite smaller fluctuations in the compression rates, it is obvious that the learning factor, rounded to one digit, is 1.0 for both methods. This is not surprising since both methods do not form dependencies between the simulation results. The compression rates of the PPCA Offline variants are significantly higher than the compression rates of FEMZIP. For the best combination - the PPCA Offline method combined with the encoder iMc 2+3D - a compression rate of 326.4 is achieved. This is 5.1 times higher than the compression rate of FEMZIP. Thus, an additional order of magnitude in compression rate is achieved by the additional consideration of an array of simulation results. The compression rates are similar for all PPCA Online methods, only the encoder zlib cuts off worse. It is interesting to note that when using the tree differences in combination with the iMc encoder, the compression rates are negatively affected. However, the effect for one simulation result is less than 1% while the difference is significantly higher for all 50 simulation results. It is true that with an increasing number of simulation results, it is often worthwhile to take many principal components, since these only have to be saved once for all simulations. In such cases the residual is very small. An application of the tree differences is not necessary in this case and can be, as the results show, counterproductive. Already in the case that only one simulation data set is available, the PPCA Offline method performs better by a factor of up to 2 than the state-of-the-art tool FEMZIP. This is partly due to the fact that the number of time steps in this model is high, which favors the PPCA method.

In the Silverado model there is a 1D component that combines all welding spots in the front of the car. Each weld spot is a single connected component in the sense of graph theory. The iMc encoder was developed for connected 2D and 3D parts, so that such a 1D part is the worst case. Looking at the compression rates of the PPCA Offline method depending on the applied encoder, it can be seen that the Rice encoder achieves a compression factor of 42.3, the encoder zlib 38.5 and the iMc encoder only a compression factor of 17.0. Since such elements are typical for a vehicle simulations, we investigate the iMc encoder iMc 2+3D, which uses the Rice encoder for 1D components.

It is interesting to note that all iMc encoders achieve very similar compression factors for a simulation result and achieve a 8% better compression factor than the Rice encoder and almost 22% better compression factor than the zlib encoder. With an increasing number of simulation results, the Rice encoder catches up with the iMc encoder, so that with 50 models there is only a 5% difference in the compression factor. The result of the PPCA Offline method combined with the encoder zlib behaves differently. Taking the PPCA Offline method with the iMc 2+3D encoder as reference, the difference in compression factors for the complete data set is 36% compared to 22% for one data set.

For the PPCA Online case the results are different. The compression factors given in the evaluation for the PPCA Online method were generated as follows. First, the PPCA Offline Method is applied to one simulation result. Then all further simulation results are added one after the other using a PPCA Online application. In this case there are 49 PPCA Online applications. The best result in this case is achieved by the Rice encoder, which has a significantly higher learning factor than the iMc encoder variants. The compression factor for the PPCA Online method in combination with the Rice encoder is 4% higher than the best iMc variant. For the zlib encoder, the learning factor is smaller than for the Rice encoder, but larger than the learning factor of the iMc encoder. The difference between the compression rate of the zlib compared to the iMc encoder is 17% for the entire dataset, compared to 22% difference based on one model. The reason for this can be illustrated by the 2D element components driver's door and hood, see Section 6.2.1. For both components and all encoders it is optimal to use principal components in the PPCA Offline case for the entire data set. The number of principal components between the Rice, zlib and iMc encoders differs by a maximum of 20%. In all cases, Rice and zlib encoders select at least as many principal components for prediction as the iMc encoder variants. This is because the iMc encoder can efficiently compress certain dependencies in the residuals, whereas the Rice and zlib encoders primarily benefit from small values. In the PPCA Online application, the iMc encoder's ability to take fewer principal components is counterproductive. For example, no principal components are taken for the x- and z-coordinates of the part hood for the iMc encoder and the iMc T encoder, see Table 6.13 and Figure 6.10a. Although the compression rates for the iMc encoder are high for the first simulation result, without storing principal components there is no learning effect that could improve the compression of subsequent simulation results. For the y coordinate of the driver's door component, no principal components are selected for compression until the 20th model, see Figure 6.10b. For the Rice encoder, principal components are used for compression from the second simulation onwards, and for the zlib component from the first result onwards. Since the compression with time differences as fall-back solution works so well for the iMc encoder, we have many cases where the compression rates achieved when using the fall-back solution are better than the PPCA method on one simulation result. Since the optimization process itself cannot estimate whether simulation results will be added later, it selects the number of principal components that is optimal for this case. If we are willing to sacrifice compression rate versus better PPCA Online performance, we can modify the final comparison between the best possible PPCA compression rate and the compression rate of the fall-back solution so that the PPCA methods are preferred, e.g. by multiplying the compression rate of the fall-back solution by a value less than 1 before the comparison. Tables 6.13 and 6.14 show very well the relationship between the compression rate and the decision whether to use principal

Table 6.12: *Compression factors for the coordinates of data set Silverado applying FEMZIP-L4 and our PPCA method in Offline and Online mode. The uncompressed size is 28,659,569,600 bytes in each coordinate direction.*

Variable		x	y	z
Assigned to		node	node	node
FEMZIP-L4		14.8	16.9	15.0
PPCA Online	zlib	60.5	63.8	56.3
	Rice	65.8	74.9	63.4
	iMc T	59.3	69.9	54.4
	iMc	60.9	73.4	55.7
	iMc 2+3D T	60.0	70.7	55.0
	iMc 2+3D	61.4	74.1	56.1
PPCA Offline	zlib	84.2	88.7	79.4
	Rice	102.1	108.2	95.1
	iMc T	99.0	104.8	92.5
	iMc	109.5	116.6	102.5
	iMc 2+3D T	106.8	112.9	99.5
	iMc 2+3D	111.7	118.7	104.4

Table 6.13: *Number of selected principal components and compression factors for the part hood of the Silverado data set applying our PPCA method in Offline and Online mode. The uncompressed size is 419,580,800 bytes in each coordinate direction.*

Variable		Compression factors			Number of PCs		
		x	y	z	x	y	z
PPCA Online	zlib	25.4	69.1	18.1	381	170	363
	Rice	28.0	86.0	13.0	404	181	0
	iMc	18.9	83.8	15.5	0	125	0
	iMc T	19.4	77.5	15.5	0	125	0
PPCA Offline	zlib	40.6	93.7	33.3	383	153	483
	Rice	44.9	119.2	36.3	403	153	513
	iMc	49.3	128.7	39.6	313	123	393
	iMc T	47.0	121.4	37.7	323	133	403

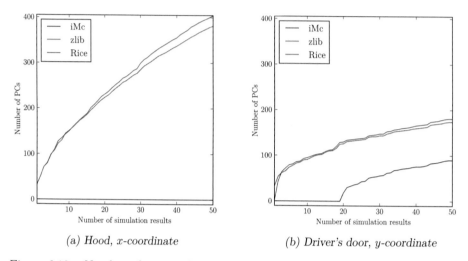

(a) Hood, x-coordinate   (b) Driver's door, y-coordinate

Figure 6.10:   Number of principal components used for PPCA prediction in respect to the number of compressed simulation results when adding one simulation result after the other with the PPCA Online method.

components for compression. The results of the PPCA Offline procedure show that for all listed cases the principal components used for compression are globally optimal. In the case that no principal components were selected, the compression rates are significantly worse than in the case where the optimization process selected principal components. The z-coordinate of the hood component is such an example. Only for the encoder zlib principal components are used for compression in PPCA Online mode. In contrast to all other cases, the zlib performs best for this specific case. For the x-coordinate no principal components are applied in the PPCA Online procedure when using the iMc encoder with and without tree differences, and thus performs significantly worse compared to the Rice encoder and the zlib encoder.

The compression rates of the coordinates in both the PPCA Online and the PPCA Offline cases are significantly better than the compression rates achieved by FEMZIP, see Table 6.12. The PPCA Online method, combined with the Rice encoder, achieves compression rates that are on average 4.4 times better than the compression factors of FEMZIP. For the PPCA Offline case, the iMc 2+3D encoder achieves a compression factor 7.2 times better than FEMZIP. The total compression rate behaves analogously to the compression rates of the coordinates for two reasons. First, the coordinates are the only node variables that have not been quantized to zero. Second, the compression rate on all element variables using the Rice encoder is very efficient, see Tables 6.15 and 6.16, or the datasets are negligibly small.

FEMZIP also performs well on the 2D and 3D element variables, see Table 6.15 and 6.16. For all examined variables, the Rice encoder in combination with PPCA provides a higher compression factor than PPCA in combination with the encoder zlib. This finding supports our approach to use the Rice Encoder as a replacement for the iMc encoder in

Table 6.14: *Compression factors and number of principal components used for compression of the coordinates for the part driver's door of the Silverado data set applying our PPCA method in Offline and Online mode. The uncompressed size is 254,600,000 bytes in each coordinate direction.*

		Compression factors			Number of PCs		
	Variable	x	y	z	x	y	z
PPCA Online	zlib	163.1	34.3	90.6	39	174	77
	Rice	265.9	41.2	119.6	42	181	81
	iMc TreeDiffs	223.9	26.8	100.7	33	90	55
	iMc	248.9	27.4	110.3	33	90	55
PPCA Offline	zlib	233.6	63.4	138.1	33	183	73
	Rice	416.5	74.4	192.5	38	183	73
	iMc	474.3	82.3	209.2	30	153	63
	TreeDiffs	443.0	78.5	198.6	30	153	63

Table 6.15: *Original and compressed sizes in bytes for 2D element variables of data set Silverado applying FEMZIP-L4 and our PPCA method in Offline and Online mode. The effective plastic strain is a three dimensional variable.*

Variable		effective plastic strain	internal energy	thickness
Assigned to		2D element	2D element	2D element
Uncompressed size		79,635,019,200	26,545,006,400	26,545,006,400
FEMZIP-L4		229.8	661.6	1,019.1
PPCA Online	zlib	208.5	383.6	512.4
	Rice	327.7	1,720.9	5,341.5
PPCA Offline	zlib	210.8	441.6	512.4
	Rice	327.7	1,722.0	5,357.7

Table 6.16: *Original and compressed sizes in bytes for 3D element variables of data set Silverado applying FEMZIP-L4 and our PPCA method in Offline and Online mode. The variable sigma has six dimensions, see Section 2.1.*

Variable		effective plastic strain	sigma
Assigned to		3D element	3D element
Uncompressed size		1,620,107,200	9,720,643,200
FEMZIP-L4		30.7	18.5
PPCA Online	zlib	29.8	46.6
	Rice	40.7	49.8
PPCA Offline	zlib	31.4	67.2
	Rice	40.7	83.0

Table 6.17: *Original and compressed file size in bytes applying FEMZIP-L4, 7zip, and our PPCA method in Offline and Online mode on data set OEM 1, see Section 6.2.2. The PPCA method is combined with the encoder zlib, Rice, and iMc for the compression of the residual of the PPCA prediction.*

		Size		Compression factor		Learning factor
Number of results		1	24	1	24	24 vs. 1
Original		4,920,557,568	118,093,381,632	1.0	1.0	1.0
FEMZIP-L4		248,041,472	5,957,046,272	19.8	19.8	1.0
7z		1,175,257,444	28,202,870,072	4.2	4.2	1.0
PPCA Offline	zlib	165,228,687	1,801,290,449	29.8	65.6	2.2
	Rice	148,694,728	1,479,413,517	33.1	79.8	2.4
	iMc	145,650,217	1,533,758,586	33.8	77.0	2.3
	iMc T	144,810,811	1,540,168,274	34.0	76.7	2.3
	iMc 2+3D	136,902,568	1,412,767,445	35.9	83.6	2.3
	iMc 2+3D T	144,361,610	1,523,548,272	34.1	77.5	2.3
PPCA Online	zlib	-	2,189,795,386	-	53.9	1.8
	Rice	-	1,841,831,003	-	64.1	1.9
	iMc	-	2,163,976,230	-	54.6	1.6
	iMc T	-	2,155,149,075	-	54.8	1.6
	iMc 2+3D	-	1,856,653,343	-	63.6	1.8
	iMc 2+3D T	-	2,139,033,562	-	55.2	1.6

unsupported cases.

6.4.2 OEM 1

The OEM 1 data set consists of 24 models with a size of approx. 4.6 GB each. Since there are no geometrical changes except for a new meshing of connection components, all data sets are of the same size. The total data set has a size of 110.0 GB. The results based on the compression rates based on a model and the complete data set can be found in Table 6.17. The data set can be reduced by a factor of 4.2 using 7z compression, which has no understanding of the data format but interprets it merely as a stream. The high compression rate for a universal encoder is achieved by first compressing the simulation results using the parameter file from Table C.2 with FEMZIP and then decompressing them again so that the effect of quantization is included in the compression rates. The state-of-the-art tool FEMZIP combines quantization with data understanding and can thus predict data and compress only the residual of the prediction instead of the original value. This increases the compression rate by a further factor of 4.7 compared to 7z, so that a compression rate of 19.8 is achieved. FEMZIP is able to exploit dependencies within a simulation result, but not between several simulations. If all 24 simulations are already available, they can be compressed using a PPCA Offline application. Using the best combination of encoders for the residuals for this use case - iMc 2+3D - an improvement by a factor of 4.2 compared to FEMZIP is achieved, resulting in a total compression rate of 83.6. Thus, an improvement of the compression rate by a further order of magnitude is achieved. The PPCA methods use a dimensionality reduction method, but the reconstruction is only used as a prediction, so the residual that has the initial dimension must be stored. Due to the high data volume of the residuals, the compression method used is crucial in determining whether the PPCA methods achieve higher compression factors than alternative approaches such as FEMZIP. Therefore, we have tested several compression methods that differ primarily in the use of the encoder. This investigation is justified, as can be seen from the results of the PPCA Offline Method applied to all 24 models of the OEM 1 data set. Using the zlib encoder results in a 27.5% larger compressed file compared to using the iMc 2+3D encoder. However, there is a much smaller difference between the results of the Rice encoder and the iMc variants. This is not surprising, as the iMc encoder for OEM 1 data is only applied to the coordinates as the only node-based variable. The Rice encoder is used for all other time-dependent variables. If the iMc encoder is applied to the coordinates of all parts in combination with tree differences, the compressed file is 9% larger than for the best tested case. Using the Rice encoder results in 4.7% larger results. In addition, an application of the tree differences in this case is counterproductive and has a negative effect on the compression factor. This is much smaller than 7.8% for iMc 2+3D with 0.4% overhead in the case that the residuals of all components are compressed with the iMc encoder. This suggests that the tree differences have a strong positive effect on the compression efficiency of 1D elements. Nevertheless, the iMc encoder provides lower compression rates than the Rice encoder for the 1D component node variable data.

In the case that only a simulation result is considered, the PPCA methods provide better compression rates than the state-of-the-art tool FEMZIP. One reason for this behavior is the high number of time steps of the model. Compared to the best encoder combination iMc 2+3D, the FEMZIP compressed file is 81.2% larger. The difference between the best performing encoder combination and the result of the worst performing encoder, zlib, is less than the difference for the complete data set based on a simulation result with a 20% larger compressed file. One reason for this is that the one-time saving of the grid connectivity in the case of one simulation result has a greater influence than in the case of 24 results. In the case of one simulation result as well as in the case of 24 results, the mesh description amounts to 11.5 MB in the compressed file. However, this explains only 2.5% of the difference between zlib results and iMc 2+3D results. The remaining 5% are due to the fact that the iMc benefits from the larger amount of data and it is worthwhile for the Rice encoder to store many principal components, so that the residual contains only small absolute values. This behavior can also be observed with the difference between both encoders iMc 2+3D and Rice. If the difference is only 4.7% for 24 models, it is 8.6% for one model.

As mentioned in the introduction of this section, the question of whether PPCA procedures benefit from the existence of many similar simulation results should be clarified. This can clearly be answered with yes, regardless of which encoder is used for the residual of the PPCA prediction. The learning factor for all combinations of PPCA offline method and residual encoder is greater than 2.2, with the Rice encoder having the largest learning factor. For the PPCA Online methods, the learning factor is between 1.6 and 1.9, with the Rice encoder also having the highest learning factor. This is due to the fact that the Rice encoder benefits greatly from very small values and therefore prefers a higher number of principal components compared to the other encoders. For the compression

Table 6.18: *Original and compressed sizes in bytes for the coordinates of data set OEM 1 applying FEMZIP-L4 and our PPCA method in Offline and Online mode.*

Variable		x	y	z
Assigned to		node	node	node
Uncompressed size		24,175,241,856	24,175,241,856	24,175,241,856
FEMZIP-L4		1,614,815,200	1,548,330,824	1,573,148,960
PPCA Online	zlib	578,130,063	537,598,512	586,361,917
	Rice	538,789,881	456,791,053	507,624,312
	iMc T	645,704,204	573,404,830	597,568,919
	iMc	659,302,842	569,712,150	596,367,673
	iMc 2+3D T	639,643,635	568,555,903	592,263,145
	iMc 2+3D	543,918,986	459,026,264	515,114,897
PPCA Offline	zlib	457,381,277	424,305,952	448,253,432
	Rice	397,084,867	361,262,670	386,289,950
	iMc T	413,230,712	383,454,807	408,647,989
	iMc	419,695,701	378,634,524	400,578,331
	iMc 2+3D T	407,618,988	378,157,771	402,942,470
	iMc 2+3D	376,360,925	340,666,014	360,949,461

of the x-coordinates of the first model of the data set, using the Rice encoder, in 1854 cases principal components are taken as opposed to 1770 in the case of iMc. Since the principal components represent the basis of a learning effect, it is not surprising that the learning factor for the Rice is higher than that of the iMc encoder. Surprisingly, when using the zlib encoder for 1926 components, the principal components are determined, but the learning effect is lower than with the Rice encoder.

The iMc encoder has not yet been implemented for the element-based variables. The compression rates are already very good when using the Rice encoder, see Table 6.19. So an improved compression of the element variables would only slightly affect the overall compression rate for our benchmark data sets. The share of coordinates is already very high in the uncompressed file with over 61.4%, see Table 6.18 and 6.17. After the compression, due to the good compression rates for element variables, the share of coordinates increases to 76.3% for the encoder iMc 2+3D. In contrast, the 2D and 3D element variables account for only 12.5% of the total compressed size. A significant improvement of the overall compression rate can only be achieved by an improved compression of the coordinates.

For all element based variables the Rice encoder is at least as good as the zlib encoder. For the data set OEM 1, if a PPCA Offline application is possible, using the iMc 2+3D encoder is optimal. In case the simulation results are not available from the beginning and are compressed one by one using the PPCA Online method, the use of the Rice encoder on the residuals is optimal, but the advantage over the iMc 2+3D encoder is small. However, based on the difference of the learning factor, it can be assumed that the difference will increase with additional simulation results. Based on the results of this section, it is correct to classify the PPCA methods as a method that can benefit from the existence of several similar simulation results.

Table 6.19: *Original and compressed sizes in bytes for all 2D and 3D element variables of data set OEM 1 applying FEMZIP-L4 and our PPCA method in Offline and Online mode.*

Variable		MX PLAST e	Thickness	Effect. Plastic Strain EPLE
Assigned to		2D element	2D element	3D element
Uncompressed size		20,940,625,152	20,940,625,152	1,424,731,680
FEMZIP-L4		161,800,412	81,792,544	2,654,448
PPCA	zlib	239,207,547	88,988,644	1,480,100
Online	Rice	129,585,720	47,623,733	3,154,714
PPCA	zlib	233,746,773	78,646,554	1,480,100
Offline	Rice	129,586,402	44,266,935	3,154,714

6.4.3   OEM 2

The results of our compression test for the OEM 2 data set can be found in Table 6.20. It is noticeable that the compression rates for the PPCA methods are significantly lower than for the OEM 1 case. This is due to the fact that only one time step was saved every 2ms for this example, in contrast to 1ms for the OEM 1 example. Since the simulation period remained the same and was even shortened for four models, less than half the time steps are available for the PPCA methods compared to the OEM 1 example. The compression rates for FEMZIP, on the other hand, are slightly better than for the OEM 1 example. The universal encoder 7z performs surprisingly well and achieves a compression rate of 6.9. The quantization has a strong positive effect on this data set. Compared to FEMZIP, all PPCA Offline methods, with the exception of the residual encoder zlib, perform better than the state-of-the-art tool FEMZIP. Interestingly, in the OEM 2 data set, the iMc coding methods, which perform an application of tree differences as a pre-processing step, perform better than the iMc variants without tree differences. This means that there are correlations in the residuals that are eliminated by the tree differences.

The learning factor is significantly smaller than for the data set OEM 1. This is partially due to the fact that there are only 10 models in total, each with only 62 or 77 time steps. Thus the data matrix in the PPCA offline matrix has a dimension of $710 \times n$, where n represents the number of nodes. This is significantly less than the $3408 \times n$ in the case of the OEM 1 example. For the x-coordinate of an OEM 2 simulation result, the optimization process of the PPCA Offline method selects principal components for 4432 parts when using the iMc encoder. For the Rice encoder there are 4414 parts whose x-coordinates are predicted using principal components, for the zlib encoder 4750 parts. Like for OEM 1, the zlib encoder has the highest number of parts compressed by the principal components. However, the learning factor for the OEM2 data set is also lower for the zlib than for iMc and Rice encoders. This shows again that the zlib encoder in combination with the PPCA methods performs worse than the Rice encoder and the iMc variants. For FEMZIP the learning factor is less than 1, which is due to the fact that four models with fewer time steps were added later where the compression rates achieved by FEMZIP are worse than for the 77 time step models. Compared to FEMZIP, the PPCA offline methods show an improved compression rate by a factor of 1.5 for all iMc variants. The Rice encoder performs about 10% worse with a learning factor of 1.35 for

Table 6.20:   *Original and compressed file size in bytes applying FEMZIP-L4, 7zip, and our PPCA method in Offline and Online mode on data set OEM 2, see Section 6.2.3. The PPCA method is combined with the encoder zlib, Rice, and iMc for the compression of the residual of the PPCA prediction.*

		Size		Compression factor		Learning factor
Number of results		1	10	1	10	10 vs. 1
Uncompressed Size		17,400,948,736	160,642,232,320	1.0	1.0	1.0
FEMZIP-L4		823,574,528	7,794,929,664	21.1	20.6	1.0
7z		2,540,093,356	23,359,717,351	6.9	6.9	1.0
PPCA Offline	zlib	951,558,436	7,544,271,741	18.3	21.3	1.2
	Rice	780,211,129	5,735,843,334	22.3	28.0	1.3
	iMc	713,396,762	5,215,219,384	24.4	30.8	1.3
	iMc T	697,634,244	5,135,332,310	24.9	31.3	1.3
	iMc 2+3D	712,442,878	5,206,527,102	24.4	30.9	1.3
	iMc 2+3D T	696,851,203	5,118,906,744	25.0	31.4	1.3
PPCA Online	zlib	-	7,836,383,464	-	20.5	1.1
	Rice	-	6,188,912,762	-	26.0	1.2
	iMc	-	5,705,589,733	-	28.2	1.2
	iMc T	-	5,560,228,109	-	28.9	1.2
	iMc 2+3D	-	5,690,258,232	-	28.2	1.2
	iMc 2+3D T	-	5,545,382,765	-	29.0	1.2

this data set. The use of the zlib encoder in combination with the PPCA Offline method provides only slightly better results than the state-of-the-art tool FEMZIP. In the case of the PPCA Online method the compression factors achieved by the zlib encoder are even worse. The iMc variants achieve an average compression rate 1.4 times better than FEMZIP for use in the PPCA Online method. The Rice encoder achieves an improvement of 1.26. Thus, for the OEM 2 data set, the use of the iMc encoder is optimal for both the PPCA Offline and the PPCA Online case. Interestingly, in contrast to the OEM 1 model, the tree differences have a positive effect and achieve the best compression factors. The best iMc variant for OEM 1, iMc 2+3D applied, delivers in the PPCA offline case with 1.7% minimally worse results than using the iMc 2+3D T. In the case of the PPCA Online method, the difference is 2.6%. For the data set OEM 2, as for the data set OEM 1, a significant improvement of the compression rate can only be achieved by an improved compression of the node-based variables, especially the velocities, see Table 6.21 and 6.22.

The aim of Section 6.4 was to demonstrate that PPCA procedures are suitable for learning from the existence of several similar simulation results. This is the case for all data sets examined, which is expressed by a learning factor, see Equation (6.1), greater than 1. Interestingly, at least for models that contain an above-average number of time steps, the compression rates for a single simulation result are better than in the comparison tool FEMZIP. For the data set OEM2, similar or slightly better compression rates are achieved. This behavior is positive, since the tendency is to store more rather than fewer time steps.

Table 6.21: *Compression factors for all coordinates and velocities applying FEMZIP-L4 and our PPCA method in Offline and Online mode on data set OEM 2. The original size is 14,775,494,760 bytes per variable per coordinate direction.*

Variable		Coordinates			Velocity		
		x	y	z	x	y	z
Assigned to		node	node	node	node	node	node
FEMZIP-L4		14.3	15.7	15.0	13.9	13.2	12.9
PPCA Online	zlib	17.7	18.8	18.7	12.1	11.1	11.1
	Rice	21.8	24.1	24.0	13.4	12.3	12.7
	iMc T	24.4	26.0	26.0	15.5	14.1	14.9
	iMc	23.7	25.7	25.3	14.8	13.8	14.3
	iMc 2+3D T	24.7	26.2	26.2	15.5	14.1	14.9
	iMc 2+3D	23.9	25.9	25.4	14.8	13.8	14.4
PPCA Offline	zlib	19.4	20.9	21.0	12.3	11.3	11.3
	Rice	24.4	26.7	26.8	14.5	13.3	13.7
	iMc T	27.7	29.4	30.0	16.7	15.1	15.9
	iMc	27.6	29.6	29.8	16.1	14.8	15.5
	iMc 2+3D T	28.1	29.7	30.3	16.7	15.1	15.9
	iMc 2+3D	27.8	29.7	29.9	16.1	14.8	15.5

Table 6.22: *Compression factors for all variables assigned to 2D and 3D elements applying FEMZIP-L4 and our PPCA method in Offline and Online mode on data set OEM 2.*

Variable		elem. 1	elem. 2	inter.	thickness
Assigned to		2D el.	2D el.	2D el.	2D el.
FEMZIP-L4		60.1	60.1	47.4	239.6
PPCA Online	zlib	28.6	28.6	34.6	195.3
	Rice	69.3	69.4	57.4	373.6
PPCA Offline	zlib	28.5	28.4	34.1	203.0
	Rice	69.3	69.4	58.6	393.9

Another finding is that it is worth investigating which encoder achieves the best compression rates. For the zlib encoder there is even a single case in which the compression rate of the PPCA Online method is worse than FEMZIP, see Table 6.20. The best compression rates are achieved with the Rice encoder and the iMc Rice combination iMc 2+3D. For the PPCA Offline method, the iMc 2+3D encoder always delivers the best results. For the PPCA Online methods both the Rice encoder and the iMc 2+3D are a good choice. The combination of the PPCA methods with the zlib encoder, on the other hand, delivers the worst results of all the encoders examined for all applications. We, therefore, advise against using the zlib in combination with the PPCA methods.

It is noticeable that the iMc methods do not achieve good compression factors on 1D components. This is no surprise, since the iMc methods were developed for the compression of 2D and 3D grids. Especially with 1D components that represent welding points, connect two nodes and are distributed over the complete model, the Markov chain approach is not effective. This behavior could be observed with all data sets, so that we recommend not to compress 1D components with the iMc encoder.

The results clearly show that a significant improvement of the compression rates can only be achieved if the compression rates for the node variables can be improved.

6.5   Run times

In this section we investigate the run times of the PPCA method and how fast the PPCA methods are compared to the state-of-the-art tool FEMZIP and the universal encoder 7z. In addition, we investigate how the choice of encoder affects the run time. For the interpretation of the run times it is important to know in which workflow the methods are integrated. We currently assume that the data is available either in FEMZIP-P or in FEMZIP-L compressed format. Using the C-API libfemunzip we can access the geometry and individual variables in the compressed file for each time step. The C-API does not provide the possibility to query data for individual parts. Only the complete result vector containing e.g. the x-coordinate for all nodes can be returned. We, therefore, implemented the workflow as an out-of-core procedure, since we assume that it is not possible to keep the x-coordinates for all nodes for all time steps of all simulation results in the RAM. After being decompressed using the libfemunzip API, the data is first written to a working directory on the SSD. The decompression and writing process represents a considerable effort, e. g. 453 GB for the Silverado data set in the case of PPCA Offline. The focus of the PPCA methods and the iMc encoder is not on a fast compression but on a fast decompression. The standard application in the vehicle safety development is to read the compressed file into a postprocessor, e. g. GNS Animator [46]. For this purpose, the data must be provided via an interface, such as the libfemunzip API. We compare the decompression times of the PPCA procedures in combination with the zlib, Rice and iMc encoder variants with the decompression times of the libfemunzip. We assume that the complete simulation result will be decompressed. For quality control a decompression was realized which does not write into the native crash format but into the FEMZIP compressed format. This means that for decompression the PPCA decompression methods and the respective decoder are used and a FEMZIP compression is applied to the result. Therefore, the decompression run times are not meaningful. All run time tests were performed on a Scientific Linux 7 machine with two Intel Xeon Silver

Table 6.23:   *Achieved compression run times and decompression run times in seconds applying FEMZIP and 7z on the first simulation result of each data set. We used 8 OMP threads for the PPCA Offline compression, while 7z was applying 32 threads and FEMZIP ran single threaded.*

Data set	Compression					Decompression	
	lib-femunzip	7z	PPCA Offline			lib-femunzip	7z
			Rice	iMc 2+3D T	zlib		
Silverado	83.9	195.1	94.2	118.4	113.5	40.2	75.9
OEM 1	54.9	182.1	88.7	112.2	115.9	21.3	96.5
OEM 2	186.8	766.3	346.1	488.7	567.2	82.5	190.5

4110 processors with 2.1 GHz and 8 cores each and 128 GB DDR4 RAM. The data were read and written on a Samsung SSD 970 PRO 1TB. For the run time investigations of the PPCA methods we concentrated on the encoders zlib, Rice and iMc 2+3D T. The approach for the iMc encoder to consider only 2D and 3D components proved to be useful for all data sets in terms of compression efficiency. The iMc 2+3D T encoder with tree differences also represents an upper limit for the run times of the iMc 2+3D encoder. Table 6.23 shows the compression run times for the application of the compression tools to one simulation result from all benchmark data sets. FEMZIP has the fastest compression in all benchmarks. The throughput rate is between 85 and 90 MB per second for the data sets OEM 1 and OEM 2. FEMZIP even achieves a throughput rate of 110 MB/second for the data set Silverado. With this data set, FEMZIP, like the other compression methods, benefits from the zero blocks of the roughly quantized variables. In general, a high compression rate is accompanied by a fast compression time. The 7z compression times are slower than FEMZIP by a factor of 2.3 for the Silverado up to 4.1 for the OEM 2 data set.

All variants of the PPCA Offline method applied to one simulation result are ranked between the short run times of FEMZIP and the longer run times of 7z. The residual encoder for the PPCA Offline method has a large influence on the run time. The Rice encoder is the fastest residual encoder. This is not surprising since the Rice encoder is a symmetric encoder in contrast to the asymmetric encoders iMc and zlib. For the iMc encoder, the compression run times are slower between 25.7% for a Silverado simulation result and 41.2% for one OEM 2 simulation result. Compared to the Rice encoder, the zlib encoder has a larger variation from 20.0% for the Silverado up to 63.9% for data set OEM 2. The differences between the encoders are associated with the compression factors as follows: The higher the compression factor, the faster the data can be encoded. In the case of fast encoding and a constant run time of the other steps of the compression process, it holds that the influence of the encoder on the run time is lower than in case of slow encoding. In Section 5.2 we have stated that the run time of the iMc encoder depends on the number of nonzero elements in the transition matrix and the maximum number of nonzero elements in a column. For high entropy data, both numbers are big, which explains the high run time. For the Silverado and OEM 1 examples, the compression run times using the iMc 2+3D T encoder are similar to the zlib compression run times. The OEM 2 example shows a difference of 16.1% between the application of iMc 2+3D T and zlib encoder. The OEM 2 data set is also the example in which the compression rates for

Table 6.24: *Compression run times in seconds applying our PPCA method in Offline and Online mode on the full data sets. The PPCA method is combined with the encoder zlib, Rice, and the combination of iMc for 2D and 3D parts together with tree prediction and Rice for 1D parts for the compression of the residual of the PPCA prediction. We used 8 OMP threads for the PPCA Offline and PPCA Online compression.*

Model	PPCA Offline			PPCA Online		
	iMc 2+3D T	Rice	zlib	iMc 2+3D T	Rice	zlib
Silverado	37,125.8	36,245.1	37,335.1	5,764.2	4,958.2	5,944.2
OEM 1	11,489.0	11,025.6	12,425.2	2,429.7	2,240.4	2,901.0
OEM 2	8,580.9	7,056.6	10,099.5	3,972.0	2,959.6	5,348.3

the node variables differ most between the use of the iMc encoder and the zlib, see Table 6.20.

The Rice encoder provides the best run time results for the PPCA methods. Since our focus is not on fast compression, the overhead is acceptable compared to FEMZIP with 12.3% for the Silverado, 61.5% for OEM 1 and 85% for OEM 2 data sets. For the iMc 2+3D T and the zlib encoder, the run times are significantly higher. At least all encoders achieve a reasonable throughput of 30 MB per second. In the case of the Silverado data set, both, iMc 2+3D T and zlib, reach a throughput of about 80 MB per second.

The time for decompression and writing the data to SSD is also indicated in the results for FEMZIP and 7z in Table 6.23. Similar to the compression case, FEMZIP as a specialized software tool is significantly faster than 7z. The difference is between a factor of 1.9 for one Silverado result file and up to 4.5 for one simulation result of the data set OEM 1. Thus the specialized tool FEMZIP dominates the universal encoder in compression and decompression speed.

We will now look at the complete array of simulation results instead of one data set, see Table 6.24. For the PPCA Online methods, the situation is similar to that for the compression of a single simulation result. If we consider the total run time of the PPCA Online methods and divide them by the number of simulation results in each data set, we get an average run time per model. For the encoder zlib and the Rice encoder, the average run times for the Silverado data set are 99.2 and 118.9 seconds, respectively. This is about 5% worse than for a single application of the PPCA Offline Step. The same is valid for the data set OEM 1. For the data set OEM 2, this calculation is not meaningful, as the PPCA Offline Method was executed on a model with 77 time steps and in the further course of the PPCA Online benchmark 4 models are added that contain only 62 time steps. In contrast to zlib and Rice encoders, the iMc encoder benefits from the fact that more and more simulation results are available in the course of the benchmark. The run time advantage is 10.8% for data set OEM 1 and 3% for the Silverado data set. This behavior can again be attributed to a strong correlation between the run time of the iMc encoder and the property of how well a data set can be compressed. Applying the PPCA Offline method to all simulation results of a data set, the PPCA run times are high. Compared to the PPCA Online application, they differ up to a factor of 7.2 for the Rice encoder in the Silverado example. The run time differences, which are up to 85% for the PPCA Online methods, are significantly smaller for the PPCA Offline method. This is due to the fact that the eigendecomposition and the matrix computations for large

Table 6.25: *Share of eigendecomposition and matrix computations on the complete run time for our PPCA method in Offline and Online mode on the full OEM 1 data set.*

	iMc 2+3D T	Rice	zlib
PPCA Online	28.14%	36.99%	27.47%
PPCA Offline	60.90%	66.00%	54.47%

Table 6.26: *Share of the optimization process on the complete run time for our PPCA method in Offline and Online mode on the full OEM 1 data set.*

	iMc 2+3D T	Rice	zlib
PPCA Online	43.42%	36.40%	45.17%
PPCA Offline	31.52%	26.39%	37.28%

components are much more complex in the PPCA Offline case than in the PPCA Online case. We perform the eigendecomposition either on the covariance matrix or on the Gram matrix, depending on which matrix has the smaller dimension, see Section 4.3.2. For the Silverado data set, the number of time steps of all simulation results is 7600. For the parts hood and driver's door, which each contain more than 7600 nodes, see Table 6.3, this leads to an eigenvalue decomposition of a 7600 x 7600 matrix instead of 50 decompositions of a 152 x 152 matrix. Due to the higher matrix dimension, the optimization process has to work longer and more sampling points need to be evaluated until the optimization has found the best number of principal components. Therefore the absolute difference between the fastest encoder Rice, the second fastest encoder iMc and the slowest encoder zlib increases. However, the relative difference of the overall run time is reduced by the fact that eigendecomposition is the time dominating factor in the PPCA Offline method.

Table 6.25 shows the share of the eigendecomposition and matrix calculations on the total run time for the data set OEM 1. It can clearly be seen that the matrix calculations and the eigendecomposition for the PPCA Offline case make up the largest part of the total run time. For the PPCA Online case optimization is the dominating factor or, as for the Rice encoder, it is balanced with the combination of eigendecomposition and matrix computations, see Table 6.26. The share of optimization in the total run time is large for our PPCA methods. Therefore, it is very important to use a residual encoder that can quickly determine the achieved compression rates in the optimization process. For the iMc encoder, we use an entropy encoder and calculate the resulting compressed size from the entropy. This reduces the effort considerably. Although the iMc encoder is a two-pass encoder, it only needs one pass to determine the statistics. As a result, optimization for the iMc encoder is faster than optimization for the zlib encoder. The Rice encoder, however, is so fast that full encoding takes less time than building the iMc encoder's statistics.

The run time of approximately 10h for a PPCA Offline application on the entire Silverado data set is significantly shorter than the run time used to simulate the data. In contrast, the PPCA Online method in combination with the Rice Encoder would be ready after 82 minutes, i. e. more than 7 times faster. One way to reduce the run time is to determine fewer eigenvectors during the eigendecomposition. However, this can limit the optimization, since it is only executed with the calculated principal components.

Table 6.27: *Decompression to RAM run times in seconds applying on FEMZIP com-pressed, PPCA compressed version of the first simulation result of each data set. The PPCA method is combined with the encoder zlib, Rice, and the combination of iMc for 2D and 3D parts together with tree prediction and Rice for 1D parts for the compres-sion of the residual of the PPCA prediction. We used 16 OMP threads for the PPCA decompression.*

Model	FEMZIP	PPCA		
		iMc	Rice	zlib
Silverado	26.2	27.1	26.2	28.9
OEM 1	16.0	18.4	16.1	20.6
OEM 2	65.1	77.8	60.6	75.7

Therefore, the restriction should be chosen carefully.

More important than the compression speed, is a fast decompression into the memory. The engineer has to wait at least the decompression time until he can analyze the model. We compare the results with the femunziplib API used in the PPCA compression tools as reading routines and summarize them in Table 6.27. The decompression of the PPCA procedures is mathematically the same for PPCA Online and PPCA Offline methods. The main difference is that for the PPCA Online case the principal components can be distributed over several files. Since the benchmarks were executed several times in a row, one can assume that all files are in RAM. Thus the difference in accessing either one database in the PPCA Offline case or a maximum of 50 databases in the PPCA Online files does not represent a disadvantage for our PPCA Online benchmarks. The decompression of the PPCA procedures is fast and in the same order of magnitude as the decompression times of FEMZIP. For the fastest decoder - Rice - the run times for the data sets Silverado and OEM 2 are almost identical to the run times of FEMZIP. For the data set OEM 1, the run times are even faster than those of FEMZIP. The iMc decoder and the zlib decoder are slower than the femunziplib API and the Rice encoder. The difference is especially big for the OEM 2 model. The OEM 2 model has several very large parts that represent the barrier of the ODB crash. For these parts the Rice encoder is much faster than the iMc encoder. For a fast postprocessing, the PPCA methods in combination with the Rice encoder and the femunziplib API are equally recommended based on the results of Table 6.27. There is no direct plugin for the 7z universal encoder, so that in this case the standard decompression time must be considered, which is 3-4 times slower than the FEMZIP and PPCA Rice decompression times, see Table 6.23 and Table 6.27.

6.6   Application fields

We see two different applications for our PPCA method. Due to its high compression rates, the PPCA Offline Method is suitable for long-term archiving. It is legally required that certain simulation results of a vehicle development have to be kept for 10 years, cp. Section 2.2. Since the data must be stored fail-safe, the storage space is expensive over this period. Therefore an investment of a large compression run time can be useful to reduce the file sizes. In addition, decompression speed plays a subordinate role in this context.

The PPCA Online method, on the other hand, fits very well into an SDMS, see Remark 2.5. Usually, many similar simulations are generated within a vehicle development. Due to the long run time of the simulations, they are usually completed at different times. Since the results are several GB in size, one would like to avoid keeping uncompressed simulation results. The goal is to compress each result directly after its simulation. This is possible with the PPCA Online method. When using an SDMS system, the PPCA Online method has a further advantage, if the simulation is executed on a cluster and the user wants to analyze the data on his local workstation. For such workflows, the data must be transferred from the cluster to the workstation. With large simulation results, this can be time-consuming. After the initial transfer of the database db file, only a small additional database file adb and a simulation-specific file ssf have to be transferred for the reconstruction of a simulation result that has been compressed using the PPCA Online method and the database file db. This application benefits from the separation of the compressed data into database files $(a)db$ and a simulation-specific file ssf per simulation result. The advantage of a database that has already been transferred can be determined using the delta factor which will be defined as follows. Since FEMZIP is the state-of-the-art tool in data compression of crash simulations, we use the FEMZIP "-L4" compression as baseline to determine the advantage of the PPCA Online method. Let m be the number of simulation results, $CS_{FEMZIP}(sim_1), ..., CS_{FEMZIP}(sim_m)$ be the FEMZIP compressed size for the simulation results $sim_1, ..., sim_m$. Moreover, let db be the size of the database file for an PPCA Offline application on sim_1, adb_i the additional database for simulation result sim_i for $i = 2, ..., m$, and ssf_j the simulation specific file for sim_j with $j = 1, ..., m$. Let $S(f)$ represent the size of file f. Then the delta factor is defined as

$$\delta_1 = \frac{CS(sim_1)}{S(db) + S(ssf_1)}, \qquad \delta_j = \frac{CS(sim_j)}{S(adb_j) + S(ssf_j)} \ \forall j = 2, ..., m. \qquad (6.2)$$

The results for the delta factor for the three benchmark data sets are shown in Figure 6.11. For the Silverado data set, the delta factor rises sharply up to the fourth simulation result. The delta factor fluctuates strongly in the further course. The simulation results were generated with strong variations in sheet thickness of up to 20%. This variation leads to very different crash courses, see Figure 6.1. In cases where a crash course is already represented in a database file, the delta factor is high. If the crash behavior is new, the delta factor is correspondingly smaller. Such strong variations are extraordinary, as can be seen from the significantly smoother curves of the delta factors for the two industrial data sets OEM 1 and OEM 2.

For the OEM 1 data set, it is clear that the transfer of the first data set and thus the database db file already has the greatest influence on the delta factor. We assume that the first simulation was already transfered to the workstation and the database file db still exists. If we have compressed the second simulation result using the database file of the first simulation result with PPCA Online and encoder Rice, it is sufficient to transfer the additional database adb_2 and the simulation-specific file ssf_2. In this case, the file transfer would be just one third of FEMZIP compressed file size $CS(sim_2)$. Thus, the file transfer is three times faster. The simulation results two and three still have a noticeable influence on the delta factor. After the third simulation result, the delta factor fluctuates

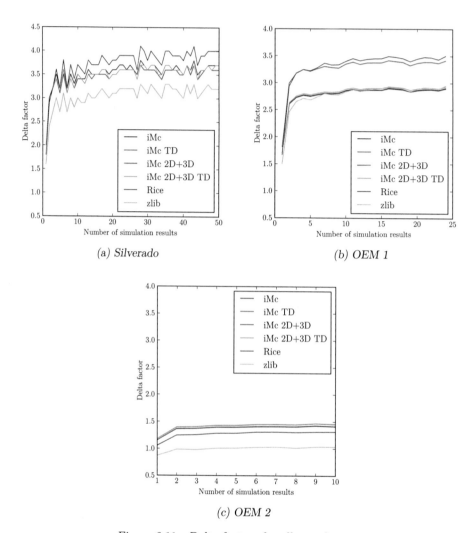

(a) *Silverado*   (b) *OEM 1*

(c) *OEM 2*

Figure 6.11: *Delta factors for all test data sets.*

only slightly. In data set OEM 2, the first transferred database has a large influence on the delta factor; the influence of all other simulation results is negligible. The delta factors are significantly smaller than for the other two data sets. This is due to the much smaller differences in compression factors between FEMZIP and the PPCA Online methods.

Conclusion:

In this chapter we have evaluated the PPCA methods introduced in Section 4.3 using three different encoders for the residual of prediction. The focus of our evaluation was on the one hand the achieved compression rates and on the other hand the run time.

The PPCA methods have several parameters that influence the run time and the compression efficiency. We calibrated these using the Silverado and OEM 1 data sets and justified our strategy for determining the number of principal components. The compression factors obtained with the PPCA methods are better than the compression factors of FEMZIP for simulation results sampled normally over time or above average when applied to a single result. However, the more efficient compression is accompanied by longer compression run time.

We distinguish two scopes of application. First, all files are already available at the time of the first PPCA compression, so that we can apply the PPCA Offline procedure, see Section 4.3.2. Second, the simulation results become available one after the other and are directly compressed with the PPCA Online procedure, see Section 4.3.3, after an initial application of the PPCA Offline procedure. As residual encoder we used the Rice encoder, see Section 3.3.3.4, the zlib, see [111], and our iMc encoder, see Section 5.1. The results show a strong dependency between the PPCA method and the encoders used. The iMc encoder achieves the best compression rates for applications within a PPCA Offline method for all benchmark data sets. The iMc encoder is not applied only to node based variables of 2D and 3D parts. For all other time-dependent data we use the Rice encoder. We call this application iMc 2+3D. The application of tree differences before iMc coding provides better results for the data set with average number of time steps and a few simulation results. For models with fine temporal resolution, however, the tree differences are counterproductive.

For the PPCA Online method, the Rice encoder achieves the best compression rates for the data sets with a high number of models and a fine temporal resolution. For the data set with only 10 simulation models and an average temporal sampling, the iMc encoder provides the best compression rates. However, both compression and decompression run times are faster with the Rice encoder. For decompression, the Rice decoder provides equal or even better run times than FEMZIP's libfemunzip API. Based on these results, we recommend using the PPCA Offline method with the iMc 2+3D encoder for long-term archiving. For the application within a simulation data management system we recommend the PPCA Online method in combination with the Rice encoder.

We have introduced two quantities in this chapter. The learning factor indicates whether a compression method benefits from the fact that several simulation data sets are available. This can be affirmed on the basis of our benchmark data sets for the PPCA methods. The delta factor reflects the advantage of building a common database for a simulation result compared to the state-of-the-art tool FEMZIP. This quantity is particularly interesting for compression methods used in an SDMS. The delta factors for the PPCA Online method justify our approach to divide the compressed results into database files (a)db and simulation-specific files ssf.

Chapter 7

Conclusion

In this chapter, we give a short summary of our proposed compression strategy for an array of simulation results. In our summary, we focus on both the predictive principal component analysis (PPCA) and induced Markov chain encoding (iMc) method from Chapters 4 and 5, respectively.
We close this chapter with an outlook and an overview of possible topics for future work.

7.1 Summary

The aim of this work was to develop a compression technique that can exploit dependencies between similar simulation results to achieve a better compression factor than compressing each simulation result separately. While being a lossy compression method, our compression techniques must be able to meet user-defined precisions. In this thesis we concentrated on crash test simulations, see Chapter 2. These are generated in large numbers as part of a vehicle development. The models simulated within a vehicle development often differ only by a few parts. In addition, several simulations with the same geometry and a parameter variation such as impact angle or sheet thickness are simulated for robustness analyses. We have designed our compression techniques to support both cases. To do this, we decompose the entire model into parts and store them in a database, see Section 4.3.1. If further simulation models contain this part, only a reference to the database is saved. This procedure eliminates redundancies in the geometry description of a set of simulation results. After the decomposition of all simulation results, it is known which component occurs in which model. A crash test simulation file consists to a large extent of time-dependent node and element variables. Therefore, the efficient compression of these variables is crucial. By decomposing the model into parts we can extract the time-dependent variables of a part for all simulation results. We interpret each time step of a model as a snapshot, and organize it as columns of a matrix. Since simulation models can behave very differently despite a strong similarity in the initial situation, an approach such as forming differences between two models at the same time step would not be effective. Therefore, we apply a PCA to the matrix with the aim of obtaining a low dimensional decomposition of the matrix. In order to save run time, we apply an eigendecomposition either to the Gram matrix or to the covariance matrix, depending on which matrix dimension is smaller. We accept numerical instabilities because we are only interested in the largest singular values and singular vectors see Section 4.2.4. When we reconstruct the matrix with the given precision, we quickly realize that this approach does not achieve dimension reduction, since almost all principal components and their coefficients have to be used. What is noticeable, however, is that the error of reconstruction across the grid is smooth and there are only small differences in values between neighboring nodes. This finding has motivated us to develop the PPCA techniques where the dimensionality reduction is used only as a prediction, see Section 4.3.2. Since the dimension of the residual of the prediction is large, we need an encoder that efficiently encodes the data. In addition, the PCA does not provide us with a spectral gap that we could use to determine the number of principal components. Therefore, we decided to use an optimization process with the objective function being the compression rate, to determine the number of principal components. This procedure leads to the fact that the compression rates of the residuals have to be determined several times. Therefore, we decided to develop our own encoder with the following features. It should have high compression rates both in the case of poorly predicted cases where no or few principal components are used, and in the case of very well predicted cases where, for example, many principal components are used. In addition, a determination of the compression rate for the optimization process should be performed quickly. Our solution is the iMc Encoder, which uses the topology of the grid as side information to achieve high compression rates for grid-based data, see Section 5.1 and [96]. The iMc encoder belongs to the category of entropy encoders, so that only the statistics have to be built up to determine the compression rate. In Section 5.1 we have introduced the iMc encoder as a practical application of our induced Markov chain model. The theoretical background of our induced Markov chain model was discussed in Section 5.4. Based on the data model we were able to show in Section 5.5 that the stationary distribution of the underlying Markov chain differs only slightly from the initial distribution of the data set. The deviation can be estimated solely on the basis of the topology of the grid. This is an important result for the efficiency of the iMc encoder as it justifies the underlying data model. In addition, we have introduced a variant for the iMc encoder that does not use a tree as a traversal and the relationship between parent and child node value to determine the statistics, but a parallelogram prediction, see Sections 3.3.2.1 and 5.7.2. Since parallelogram prediction does not only exploit dependencies between directly neighboring values, this method is categorized as a Bayesian network.

The residual encoder is crucial for the compression factor as well as for the run time, we have investigated three encoders: the iMc encoder, the Rice encoder and the zlib encoder. If all simulation models are available right away, the combination of PPCA Offline and iMc encoder applied to the node variables of the 2D and 3D components provides the best compression rates, see Section 6.4. However, the run time is high, so that this method is particularly suitable for long-term archiving of the data.

A crash test simulation takes about 8h. This means that all simulation results are usually not available at the same time. For this reason, we have adapted the idea of the PPCA Offline Method and made it an online method that can learn from new simulation results, but does not change the database of previously compressed results. Keeping the previously determined database unchanged distinguishes the method significantly from other online methods in the field of machine learning. The idea behind our approach is not to run a PCA on the original data matrix, but on the residual matrix. If the simulation result can not be described by the previously compressed principal components, the residuals are large. In this case, it is likely that our optimization process achieves the best compression rate by selecting principal components. An essential point is the reorthogonalization of the lossy compressed principal components using the amendment steps, see Section 4.3.3 and Section 6.3.2. Without applying the amendment steps, the non-orthogonal portions of the reconstructed principal components for each PPCA Online step would appear in the residuals and would have to be saved with it.

For the PPCA Online methods, the use of the Rice encoder has proved successful. This is due to the good run times of compression and decompression as well as to the fact that poorly predicted data cannot be compressed well. As a result, more principal compression components are used than, for example, in the case of the iMc encoder. Therefore, the learning factor of the PPCA Online method is higher when using the Rice encoder than when using the iMc encoder. The combination of PPCA Online method and the Rice encoder can be meaningfully applied as the data compression core of a SDMS, see Section 6.6.

For the PPCA methods there are some crucial parameters which can significantly influence the compression rate and the speed. We have examined them both theoretically and empirically on the basis of our benchmark data sets, see Section 4.3.4 and Section 6.3, respectively, and made recommendations for their selection.

With the learning factor, we have defined a quantity that indicates whether and, if so, how well a compression procedure benefits from the fact that not only a single simulation but an array of simulations is available, see Section 6.1. For all PPCA procedures based on the benchmark data sets, this quantity is greater than 1. Thus, the PPCA procedures are categorized as a machine learning procedures. Moreover, we have achieved our goal of exploiting dependencies between simulation results for successful compression and gain distinctly better compression factors than the state-of-the-art tool FEMZIP.

7.2   Outlook and future work

There are two main criteria that determine the quality of a compression. This is the run time on the one hand and the compression rate on the other. In Section 6.4 we have already shown that our PPCA compression achieves very good compression rates and is better from the first simulation model than with the state-of-the-art tool FEMZIP. However, the compression run times are worse than with FEMZIP and the decompression times for the iMc and zlib coders are worse than for the libfemunzip decompression library. In this section we list strategies and approaches that have a positive effect on the run time or compression factors of our compression technique, but have not been explored in detail in this thesis.

7.2.1   Runtime

As stated in Section 6.5, run time of compression is a crucial point that determines whether the PPCA procedures can be used meaningfully. The run time of the PPCA methods for compression is slower than with the state-of-the-art tool FEMZIP. Especially the PPCA Offline application is so slow that it only makes sense for special applications, e.g. long-term archiving. An improvement of the run time could extend the possible field of application. The optimization process for determining the number of main components accounts for about 40% of the run time in the case of PPCA Online and about 30% of the run time in the case of a PPCA Offline application. The accelerations could be achieved by an alternative approach to determining the number of principal components. Akaike's information theoretic criterion or the minimum description length criterion, see Section 4.3.4.1. This could significantly accelerate the overall run time of the PPCA compression process but will probably effect the compression rate.

In Section 6.5 we have seen that matrix calculations account for a large part of the run time of the PPCA Offline procedure. This is especially due to the super linear run times of the eigendecomposition and matrix-matrix multiplications. With the following two approaches we can reduce the matrix dimensions and thus achieve a run time advantage. First, we can limit the number of principal components that can be selected in one PPCA Offline Step. In this case, fewer eigenvectors have to be determined, the optimization process has to evaluate fewer sampling points, and the reconstructions are achieved by smaller matrix-matrix multiplications. If the constraint is too small, this may have a negative effect on the compression rate. Second, large parts can be split into several small parts. If the number of time steps is greater than the number of nodes of the disassembled part, we achieve a run time advantage for the eigendecomposition step. A disadvantage is that the sub-components have to be organized in such a way that they can be reassembled to the original component.

The iMc encoder can be combined with a RANS encoder [34] instead of an arithmetic encoder to achieve a faster execution of the encoding, see [116]. A modification could not only improve the compression but also the decompression speed.

7.2.2   Compression factor

We reached our goal to achieve higher compression factors with our PPCA methods compared to the state-of-the-art tool FEMZIP, see Section 6.4. Nevertheless, there are approaches that can further improve compression rates.

An alternative to PCA as dimension reduction method is the Independent Component Analysis (ICA) [20]. An important difference between the PCA and the ICA is that the calculated components of the ICA are generally not orthogonal. An open question is whether the amendment steps from Section 4.3.3 can be applied in such a way that the ICA can also be used as an online procedure.

For the PPCA Online method, the iMc encoder performs worse than the Rice encoder. As discussed in Section 6.4, this is because the iMc encoder can handle poorly predicted data better than the Rice encoder. As a result, fewer principal components tend to be used, which negatively affects the compression of simulation results added later. This trend can be countered by modifying the optimization process to favor a higher number of principal components. For example, this can be done by using a penalty factor that is the highest for choosing no principal components and becomes smaller to zero for the maximum number of principal components that can be selected, see Section 4.3.4.1.

For the benchmarks of Section 6.4, the iMc encoder was only applied to node-based variables. An application to element-based variables is possible if an element to element connectivity is determined. Since the compression rates for the element variables are already very good, see Section 6.4, the effect for our benchmark data sets would be small.

In Section 5.7.2, we have introduced an alternative to the iMc encoder that takes the traverse of a parallelogram prediction instead of a tree. In addition, the statistics are not determined on the basis of value transitions between father and child nodes, but on the basis of the parallelogram prediction of the value to be coded. This encoder could be used instead of the iMc encoder to compress the residuals of shell elements.

The choice of the spanning tree has an influence on the compression rate as we have seen in Section 5.6. Instead of taking a spanning tree generated by DFS, we can determine an optimal tree for iMc coding, in which case we would have to save the tree for decompression.

While further improvements in compression rates and runtime are desirable, PPCA techniques can already outperform the state-of-the-art FEMZIP tool in compression rates and decompression times. Thus, the presented methods are a competitive alternative to the commercial tool FEMZIP.

Appendix A

Basic definitions and theorems

In this chapter, we state some well known definitions and theorems that are usually known by an interested reader. The definitions and theorems are available here for completeness but would disturbflow of reading if they would be available in the previous chapters.

A.1 Probability theory

A.1.1 Definitions

Definition A.1 (powerset [29]). *Let S be a set. The powerset is the set of all subsets of S, including the empty set as well as S.*

Definition A.2 (Countable [75]). *A set S is countable if there exists a surjective function $f : \mathbb{N} \to S$.*

Definition A.3 (σ-algebra [74]). *Let X be a set and $\mathcal{P}(X)$ its powerset. A set of sets $\mathcal{A} \subset \mathcal{P}(X)$ is called σ-algebra, if it satisfies the following three properties:*

1. $X \in \mathcal{A}$

2. If $A \in \mathcal{A}$ so is its complement $A^C \in \mathcal{A}$

3. If $A_1, A_2, ... \in \mathcal{A}$ so is its countable union $\bigcup_{i=1}^{\infty} A_i \in \mathcal{A}$

Definition A.4 (measure, probability measure [35]). *Let Ω be a set and $\mathcal{A}$ a σ-algebra of Ω. A measure is a nonnegative countably additive set function. That is, a function $\mu : \mathcal{A} \to \mathbb{R}$ with*

1. $\mu(A) \geq \mu(\emptyset) = 0, \forall A \in \mathcal{A}$, and

2. if $A_i \in \mathcal{A}$ is a countable sequence of disjoint sets, then

$$\mu \left(\bigcup_i A_i \right) = \sum_i \mu(A_i).$$

If $\mu(\Omega) = 1$, we call μ a probability measure.

Definition A.5 (Probability space [74]). *A probability space is a triple $(\Omega, \Sigma, \mathbb{P})$ consisting of a arbitrary nonempty set Ω, the σ-algebra Σ and a subset of Ω called events such that:*

I

1. $\emptyset \in \Sigma$

2. *if* $A \in \Sigma$ *then also* $(\Omega \backslash A) \in \Sigma$

3. *if* $A_i \in \Sigma$ *for* $i = 1, 2, ...$ *then also* $\bigcup_i A_i \in \Sigma$

and the probability measure $\mathbb{P} : \Sigma \rightarrow [0, 1]$ *is a function on* Σ *such that:*

1. *if* $A_i \in \Sigma$ *is a countable collection of pairwise disjoint sets, then* $\mathbb{P}(\bigsqcup_i A_i) = \sum_i \mathbb{P}(A_i)$

2. $\mathbb{P}(\Omega) = 1$.

Definition A.6 (measurable space [17]). *Let* Ω *be a set and* $\mathcal{A}$ *a* σ-*algebra of subsets of* Ω. *We call* $(\Omega, \mathcal{A})$ *a measurable space.*

Definition A.7 (Laplace experiment, Laplace probability space [17]). *A Laplace experiment is a random experiment with only finitely many possible outcomes, each equally likely. The associated mathematical model is a probability space* $(\Omega, \mathcal{P}(\Omega), \mathbb{P}_N)$ *in which* Ω *has* $N \in \mathbb{N}$ *elements,* $\mathcal{P}(\Omega)$ *and* $\mathbb{P}_N$ *is the unique measure satisfying* $\mathbb{P}(\{\omega\}) = N^{-1}$ *for every* $\omega \in \Omega$. *This probability space is called Laplace probability space of order* N.

Definition A.8 (joint distribution [17]). *Let* $X_1, ..., X_n$ *be random variables with distributions* $\mathbb{P}_{X_1}, ..., \mathbb{P}_{X_n}$. *Then* $P_Y = P_{X_1 \times ... \times X_n}$ *is called joint distribution of the random variables* $X_1, ..., X_n$.

Definition A.9 (communicating class [130]). *A communicating class is an equivalence class made up of communicating states, see Definition 3.29.*

Definition A.10 (marginal distribution [35]). *Let* $X_1, ..., X_n$ *be random variables with a joint distribution* $\mathbb{P}$ *and a finite state space. Then* $\mathbb{P}_i[x] = \mathbb{P}[X_i = x]$ *for all* $i = 1, ..., n$ *are called marginal distribution.*

Definition A.11 (independent and identically distributed (i.i.d.) [35]). *Let* $X_1, X_2, ...$ *be random variables, that have the same distribution. Then they are called independent and identically distributed, or i.i.d. for short.*

Definition A.12 (stochastic process [74]). *A family of random variables* $X = (X_t, t \in \mathbb{N}_0)$ *on a probability space* $(\Omega, \mathcal{B}(\Omega), \mathbb{P})$ *with values in* $(\mathcal{A}, \mathcal{P}(\mathcal{A}))$ *is called stochastic process with time set* $\mathbb{N}_0$ *and state space* $\mathcal{A}$.

A.1.2 Statements and theorems

Proposition A.13. *Let a undirected graph* $G(V(G), E(G))$ *with the property that all walks from one node back to itself has even length.*

- *If there exists a walk from node* $n_i \in V(G)$ *to* $n_j \in V(G)$ *with odd length, every walk from* n_i *to* n_j *is of odd length.*

- *If there exists a walk from node* $n_i \in V(G)$ *to* $n_j \in V(G)$ *with even length, every walk from* n_i *to* n_j *is of even length.*

- *If you take a third node $n_c \in V(G)$ with $n_c \neq n_i$ and $n_c \neq n_j$ and a walk of odd length from n_i to n_j. Either all walks from n_c to n_i have even length and all walks from n_c to n_j have odd length or all walks from n_c to n_i have odd length and all walks from n_c to n_j have even length.*

PROOF: To the first and second point:
Assuming that there is at least one walk with odd and one with even length from n_i to n_j. Let

$$\langle n_i, n_{l_1}, ..., n_{l_s}, n_j \rangle \text{ and } \langle n_i, n_{m_1}, ..., n_{m_t}, n_j \rangle$$

be these two walks. Then the walk $\langle n_i, n_{l_1}, ..., n_{l_s}, n_j, n_{m_t}, ..., n_{m_1}, n_i \rangle$ has odd length which is a contradiction to the property of the graph that all walks from one node back to itself have even length.
To the third point:
Assuming that both walks $\langle n_c, ..., n_i \rangle$ and $\langle n_c, ..., n_j \rangle$ are both even, so there exists a even walk from n_i via n_c to n_j and so all walks from n_i to n_j are even. Assuming that both walks $\langle n_c, ..., n_i \rangle$ and $\langle n_c, ..., n_j \rangle$ are both odd, so there exists a even walk from n_i via n_c to n_j and so all walks from n_i to n_j are even. In both cases it is a contradiction that there exists a walk off odd length from n_i to n_j. □

Proof of Lemma 5.41:

PROOF: Let $s_0, ..., s_n \in \mathcal{A}$ be the states with nonzero probabilities $p_0, ..., p_n$ and $\sum_{i=0}^{n} p_i = 1$. Due to the symmetric distribution one has $c_i = -c_{n-i}$ with $p_i = p_{n-i}$ $\forall i = 0, ..., n$. At this point we have to make a case differentiation.
We define $q_i := p_i + p_{n-i}$ $\forall i = 0, ..., n/2$ if n is even and $\forall i = 0, ..., (n-1)/2$ if n is odd. The q_i correspond to $-s_i$ for $\forall i = 0, ..., n/2$ if n is even and $\forall i = 0, ..., (n-1)/2$ if n is odd.

$n + 1$ is even

$$H = -\sum_{i=0}^{n} p_i \log_2 p_i$$

$$= -\sum_{i=0}^{\frac{n-1}{2}} 2p_i \log_2 p_i$$

$$= -\left(\sum_{i=0}^{\frac{n-1}{2}} 2p_i \log_2 p_i + 1 \right) + 1$$

$$= -\sum_{i=0}^{\frac{n-1}{2}} 2p_i \left(\log_2 p_i + 1 \right) + 1$$

$$= -\sum_{i=0}^{\frac{n-1}{2}} 2p_i \log_2 2p_i + 1$$

$$= -\sum_{i=0}^{\frac{n-1}{2}} \tilde{p}_i \log_2 \tilde{p}_i + 1$$

$$= \tilde{H} + \frac{\#\text{Nonzero Entries}}{\#\text{Entries}}$$

$\underline{n+1 \text{ is odd}}$

$$H = - \sum_{i=0}^{n} p_i \log_2 p_i$$

$$= - \sum_{i=0}^{\frac{n}{2}-1} 2p_i \log_2 p_i - p_{\frac{n}{2}} \log_2 p_{\frac{n}{2}}$$

$$= - \sum_{i=0}^{\frac{n}{2}-1} 2p_i \log_2 p_i - p_{\frac{n}{2}} \log_2 p_{\frac{n}{2}} + \left(- \sum_{i=0}^{\frac{n}{2}-1} 2p_i - p_{\frac{n}{2}} + 1 \right)$$

$$= - \sum_{i=0}^{\frac{n}{2}-1} 2p_i \log_2 2p_i - p_{\frac{n}{2}} \log_2 p_{\frac{n}{2}} - p_{\frac{n}{2}} + 1$$

$$= - \sum_{i=0}^{\frac{n}{2}-1} \tilde{p} \log_2 \tilde{p}_i - \tilde{p}_{\frac{n}{2}} \log_2 \tilde{p}_{\frac{n}{2}} + \left(1 - p_{\frac{n}{2}} \right)$$

$$= \tilde{H} + \frac{\#\text{Nonzero Entries}}{\#\text{Entries}}$$

$\square$

Lemma A.14. *Let* $H = \hat{V}\hat{V}^T$ *with* $\hat{V} \in \mathbb{R}^{n \times P}$ *like in Equation (4.24) and* $P \leq n$. *Moreover, let* $D = \hat{V}^T\hat{V} \in \mathbb{R}^{P \times P}$, *it holds that*

$$spec(H) = spec(D) \cup \bigcup_{i=1}^{n-P} \{0\}. \tag{A.1}$$

Furthermore, for matrix $K = (I - H) \in \mathbb{R}^{n \times n}$ *and* $spec(D) = \{\sigma_1^2, ..., \sigma_P^2\}$, *it holds that*

$$spec(K) = \{1 - \sigma_1^2, ..., 1 - \sigma_P^2\} \cup \bigcup_{i=1}^{n-P} \{1\}. \tag{A.2}$$

PROOF: Due to Theorem 4.2, there exists a singular value decomposition of $\hat{V} = U\Sigma W^T$, with $U \in \mathbb{R}^{n \times n}$ and $W \in \mathbb{R}^{P \times P}$ orthogonal and

$$\Sigma = \begin{pmatrix} \text{diag}(\sigma_1, ..., \sigma_P) \\ 0 \end{pmatrix} \in \mathbb{R}^{n \times P},$$

see Section 4.2.2. Then

$$H = VV^T = U\Sigma W^T W \Sigma^T U^T = U \begin{pmatrix} \text{diag}(\sigma_1^2, ..., \sigma_P^2) & 0 \\ 0 & 0 \end{pmatrix} U^T$$

and

$$D = \hat{V}^T\hat{V} = W\Sigma^T U^T U\Sigma W^T = W \text{diag}(\sigma_1^2, ..., \sigma_P^2) W^T.$$

Since we found eigenvalue decompositions for matrices H and D, we can directly see the spectrum of both matrices. We have $\text{spec}(D) = \{\sigma_1^2, ..., \sigma_P^2\}$ and $\text{spec}(H) = \text{spec}(D) \cup \bigcup_{i=1}^{n-P}\{0\}$. Since all eigenvectors of H are also eigenvectors of $K = (I - H)$, we have

$$K = (I - H)$$
$$= UIU^T - U\begin{pmatrix} \text{diag}(\sigma_1^2, ..., \sigma_P^2) & 0 \\ 0 & 0 \end{pmatrix} U^T$$
$$= U\left(I - \begin{pmatrix} \text{diag}(\sigma_1^2, ..., \sigma_P^2) & 0 \\ 0 & 0 \end{pmatrix}\right) U^T$$
$$= U\left(I - \begin{pmatrix} \text{diag}(1 - \sigma_1^2, ..., 1 - \sigma_P^2) & 0 \\ 0 & I_{(n-P)\times(n-P)} \end{pmatrix}\right) U^T$$

Again, we found an eigenvalue decomposition and we can directly read the eigenvalues of matrix K as the right-hand side of Equation (A.2). □

Appendix B

Ergodicity of Markov processes

Let $(\Omega, \mathcal{B}(\Omega), \mathbb{P})$ a probability space where $\mathcal{B}(\Omega)$ is the Borel σ-algebra of subsets of Ω and $\mathbb{P}$ is a probability measure.

Definition B.1 (elementary Markov property, Markov process). *Let $X = (X_t)_{t \in \mathbb{N}_0} = (X_t, t \in \mathbb{N}_0)$ be a stochastic process with values in a countable set S. The stochastic process X has the elementary Markov property if and only if for all $n \in \mathbb{N}$ and all $t_1 < t_2 < \ldots < t_n < t$ and $i_1, \ldots, i_n, i \in S$ with $\mathbb{P}[X_{t_1} = i_1, \ldots, X_{t_n} = i_n] > 0$ the following equation holds:*

$$\mathbb{P}[X_t = i | X_{t_1} = i_1, \ldots, X_{t_n} = i_n] = \mathbb{P}[X_t = i | X_{t_n} = i_n] = \mathbb{P}_{i_n}[X_{t-t_n} = i]. \tag{B.1}$$

The stochastic process X will be called Markov chain with distribution $(\mathbb{P}_i)_{i \in S}$ on $(\Omega, \mathcal{B}(\Omega))$ if

- *for every $i \in S$, X is a stochastic process on the probability space $(\Omega, \mathcal{B}(\Omega), \mathbb{P}_i)$ with $\mathbb{P}_i[X_0 = i] = 1$.*

- *the mapping $\kappa : S \times \mathcal{B}(S)^{\otimes \mathbb{N}_0} \to [0, 1], (i, B) \mapsto \mathbb{P}_i[X \in B]$ is a stochastic kernel (i.e. the mapping $B \mapsto \mathbb{P}_i[X \in B]$ is a probability measure).*

- *the weak Markov property is true: For all $A \in \mathcal{B}(S)$, all $i \in S$ and each combination of $s, t \in \mathbb{N}_0$ it holds: $\mathbb{P}_i[X_{t+s} \in A] = \kappa_t(X_s, A)$ $\mathbb{P}_i$-almost sure (a.s.). For all $t \in \mathbb{N}_0$, $i \in S$ and $A \in \mathcal{B}(S)$ the stochastic kernel $\kappa_t : E \times \mathcal{B}(S) \to [0, 1]$ of the transition probabilities of X for the time difference t is defined as:*

$$\kappa_t(i, A) := \kappa\left(i, \left\{ y \in S^{\mathbb{N}_0} : y(t) \in A \right\}\right) = \mathbb{P}_i[X_t \in A]$$

The set of κ_n will be called family of n-step transition probabilities.

In our case we restrict the countable set S to be finite and use $\mathcal{P}(S)$ as $\mathcal{B}(S)$, see also Remark 3.6. In the context of a finite state space S and discrete time $t \in \mathbb{N}_0$ the demands on a stochastic process to be a Markov chain reduces to the Markov property if there exists a transition probability matrix which is a stochastic matrix and a initial distribution ν on S.

In this case the κ_t are the rows of the transition probability matrix to the power of $t \in \mathbb{N}_0$. The probability $\mathbb{P}_i[X_t \in A]$, $i \in S$, is the sum of the columns of the row vector $\kappa_t(i, \cdot)$ corresponding to the elementary events in $\mathcal{P}(S)$.

Remark B.2. *The weak Markov property corresponds to the elementary Markov property together with time homogeneity, see Definition 3.23, of the Markov chain [74].*

Definition B.3 (Polish space [74]). *A Polish space is a seperable, completely metrizable topological space.*

Definition B.4 (Markovian semigroup [74]). *Let S be a polish space. A family $(\kappa_t : t \in \mathbb{N}_0)$ of stochastic kernels is called a semigroup of stochastic kernels or Markovian semigroup if they fulfill the Chapman-Kolmogorovian equation:*

$$\kappa_s \cdot \kappa_t = \kappa_{s+t} \quad \forall s, t \in \mathbb{N}_0 \tag{B.2}$$

Theorem B.5 ([74]). *Let $(\kappa_t)_{t \in \mathbb{N}_0}$ be a Markovian semigroup of stochastic kernels from S to S. Then there exists a measurable space $(S, \mathcal{P}(S))$ and a Markov chain $((X_t)_{t \in \mathbb{N}_0}, (\mathbb{P}_i)_{i \in S})$ on $(S, \mathcal{P}(S))$ with transition probabilities*

$$\mathbb{P}_i[X_t \in A] = \kappa_t(x, A) \quad \forall i \in S, A \in \mathcal{P}(S), t \in \mathbb{N}_0. \tag{B.3}$$

Conversely for every Markov chain X the equation (B.3) defines a semigroup of stochastic kernels. By (B.3) the finite distributions of X are determined uniquely.

Let a measurable transformation $T : \Omega \to \Omega$ be given, that plays the role of a time shift [25].

Definition B.6 (invariant [74]). *An event $A \in \mathcal{P}(S)$ is called invariant regarding T, if $T^{-1}A = A$ almost everywhere regarding $\mathbb{P}$. The σ-algebra of the invariant events will be denoted by*

$$\mathcal{I} = \{A \in \mathcal{P}(S) : T^{-1}(A) = A\}.$$

Definition B.7 (measure preserving transformation, ergodicity of a dynamical system). *A transformation T is called measure preserving if*

$$\mathbb{P}[T^{-1}(A)] = \mathbb{P}[A] \quad \forall A \in \mathcal{P}(S).$$

In this case the quadruplet $(\Omega, \mathcal{B}(\Omega), \mathbb{P}, T)$ is a measure-preserving dynamical system. If T is measure preserving and the σ-algebra of the invariant events $\mathcal{I}$ is $\mathbb{P}$-trivial (i.e. $\mathbb{P}[\mathcal{I}] = 0$ or $\mathbb{P}[\mathcal{I}] = 1$), $(\Omega, \mathcal{B}(\Omega), \mathbb{P}, T)$ will be called ergodic.

Theorem B.8 (Birkhoff). *If T is ergodic it holds:*

$$\frac{1}{n} \sum_{k=0}^{n-1} X_k \xrightarrow{n \to \infty} \mathbb{E}[X_0] \quad \mathbb{P} - a.s. \tag{B.4}$$

This theorem states that the time average of one sequence of events is the same as the ensemble average. This is known as the asymptotic equipartition property (AEP) which is a consequence of the weak law of large numbers [25].

The definition of ergodicity for dynamical systems is also unconsistently used but therefore, at least equivalently. In some publications a dynamical system is ergodic if and only if it fulfills Equation (B.4).

Lemma B.9 ([74]). *Let $X = (X_t)_{t \in \mathbb{N}_0}$ be a stochastic process with values in a polish space S. Let X be the canonical process on the probability space $(\Omega, \mathcal{B}(\Omega), \mathbb{P})$. Define the shift*

$$T : \Omega \to \Omega, \quad (\omega_n)_{n \in \mathbb{N}_0} \mapsto (\omega_{n+1})_{n \in \mathbb{N}_0}$$

with $X_n(\omega) = X_0(T^n(\omega))$ and T measurable. So X is stationary if and only if $(\Omega, \mathcal{B}(\Omega), \mathbb{P}, T)$ is a measure-preserving dynamical system.

Theorem B.10 ([74]). *The canonical process X on $(S^{\mathbb{N}_0}, \mathcal{P}(S)^{\otimes \mathbb{N}_0})$ is a Markov chain with transition probability matrix P regarding to the distribution $(\mathbb{P}_i)_{i \in S}$. Especially for every stochastic matrix P there is exactly one discrete Markov chain X with transition probabilities P.*

Definition B.11 (ergodicity of a stochastic process [74]). *The stochastic process X of Lemma B.9 will be called ergodic, if the dynamical system $(\Omega, \mathcal{B}(\Omega), \mathbb{P}, T)$ is ergodic.*

Definition B.12 (mixing [74]). *A measure preserving dynamical system $(\Omega, \mathcal{B}(\Omega), \mathbb{P}, T)$ is called mixing if*

$$\lim_{n \to \infty} \mathbb{P}[A \cap T^{-n}(B)] = \mathbb{P}[A]\mathbb{P}[B] \quad \forall A, B \in S.$$

Theorem B.13. *Let X be an irreducible, positive recurrent Markov chain on the countable set S with invariant distribution π and $\mathbb{P}_\pi = \sum_{i \in S} \pi_i \mathbb{P}_i$. It holds*

- *X is ergodic on $(\Omega, \mathcal{B}(\Omega), \mathbb{P}_\pi)$.*

- *X is mixing if and only if X is aperiodic, see Definition 3.37.*

Theorem B.14. *The dynamical system $(\Omega, \mathcal{B}(\Omega), \mathbb{P}, T)$ is ergodic if and only if for all $A, B \in mathcalA$ it holds:*

$$\lim_{n \to \infty} \frac{1}{n} \sum_{k=0}^{n-1} \mathbb{P}[A \cap T^{-n}(B)] = \mathbb{P}[A]\mathbb{P}[B].$$

The classical definition of ergodicity in the context of dynamical systems is lossy speaking that the average behavior over time and over the state space respectively is identical with probability one.

Appendix C

Parameter files

Table C.1: *Precisions file applied on Chrysler Silverado (LS-DYNA) simulation result set, see Section 6.2.1*

Node values: precision	
coordinates	: 0.1000
velocities	: 0.1000E+18
accelerations	: 0.1000E+18
Shell values: precision	
effective_plastic_strain	: 0.1000E-02
thickness	: 0.1000E-01
element_dependent_variable_1	: 0.1000E+21
element_dependent_variable_2	: 0.1000E+21
internal_energy	: 10.00
Thick shell values: precision	
sigma	: 0.000
effective_plastic_strain	: 0.000
epsilon	: 0.000
Solid values: precision	
sigma	: 1.000
effective_plastic_strain	: 0.1000E-02
1D-element values: precision	
axial_force	: 10.00
s_shear_resultant	: 10.00
t_shear_resultant	: 10.00
s_bending_moment	: 100.0
t_bending_moment	: 100.0
torsional_resultant	: 10.00
SPH node values: precision	
SPH_radius	: 0.000
SPH_pressure	: 0.000
SPH_stress	: 0.000
SPH_plastic_strain	: 0.000
SPH_density_of_material	: 0.000
SPH_internal_energy	: 0.000
SPH_strain	: 0.000
SPH_mass	: 0.000
Contact segment values: precision	
intfor_pressure	: 0.000
intfor_shear	: 0.000
intfor_force	: 0.000
intfor_gap	: 0.000
intfor_energy	: 0.000
intfor_time	: 0.000

Table C.2: Precisions file applied on OEM 1 (Pam-Crash) simulation result set, see Section 6.2.2

Node values: precision		
geometry	:	0.1000E-00
Shell values: precision		
MX PLAST e EPMA	:	0.1000E-02
Thickness THIC	:	0.1000E-01
Tool values: precision		
Tool variable 1	:	0.1031E-37
Solid values: precision		
Effect. Plastic Strain EPLE	:	0.1000E-02

Table C.3: Precisions file applied on OEM 2 (LS-DYNA) data set, see Section 6.2.3

Node values: precision		
coordinates	:	0.1000
velocities	:	100.0
Shell values: precision		
thickness	:	0.1000E-01
element_dependent_variable_1	:	0.1000E-02
element_dependent_variable_2	:	0.1000E-02
internal_energy	:	0.1000
Thick shell values: precision		
Solid values: precision		
sigma	:	0.000
effective_plastic_strain	:	0.000
epsilon	:	0.000
plastic_strain_tensor	:	0.000
thermal_strain_tensor	:	0.000
1D-element values: precision		
SPH node values: precision		
Contact segment values: precision		

List of Tables

List of Figures

Bibliography

[1] AHMED, N., NATARAJAN, T., AND RAO, K. R. Discrete cosine transfom. *IEEE Trans. Comput. 23*, 1 (Jan. 1974), 90–93.

[2] AKCOGLU, M., BARTHA, P., AND HA, D. *Analysis in Vector Spaces*. Wiley, 2011.

[3] ALEXA, M., AND MÜLLER, W. Representing animations by principal components. *Computer Graphics Forum 19*, 3 (August 2000), 411–418. ISSN 1067-7055.

[4] ALPAYDIN, E. *Introduction to machine learning*. MIT press, 2014.

[5] ALTAIR ENGINEERING. Radioss. `https://www.altair.com/radioss-applications/`. Accessed on 2022-01-03.

[6] AMANN, H., AND ESCHER, J. *Analysis 2. Deutsche Ausgabe*. Birkhäuser, 2005.

[7] AMJOUN, R., SONDERSHAUS, R., AND STRASSER, W. Compression of complex animated meshes. In *Advances in Computer Graphics*, T. Nishita, Q. Peng, and H.-P. Seidel, Eds., vol. 4035 of *Lecture Notes in Computer Science*. Springer Berlin Heidelberg, 2006, pp. 606–613.

[8] ANDERSON, J. *Computational Fluid Dynamics*. Computational Fluid Dynamics: The Basics with Applications. McGraw-Hill Education, 1995.

[9] BAUM, L. E., AND PETRIE, T. Statistical inference for probabilistic functions of finite state markov chains. *The annals of mathematical statistics 37*, 6 (1966), 1554–1563.

[10] BEN-GAL, I. *Bayesian Networks*. John Wiley & Sons, Ltd, 2008.

[11] BODDEN, E. Arithmetic coding. `http://www.bodden.de/legacy/arithmetic-coding/`. Accessed on 2022-01-03.

[12] BODDEN, E., CLASEN, M., AND KNEIS, J. Arithmetic coding revealed - a guided tour from theory to praxis. Tech. Rep. SABLE-TR-2007-5, Sable Research Group, School of Computer Science, McGill University, Montréal, Québec, Canada, May 2007.

[13] BOURAGO, N. G., AND KUKUDZHANOV, V. N. A review of contact algorithms. *Mechanics of Solids 40*, 1 (2005), 35–71.

[14] BOWER, A. F. *Applied mechanics of solids*. CRC press, 2009.

[15] BOYD, P. L. Nhtsa's ncap rollover resistance rating system. In *Proceedings of the 19th International Technical Conference on the Enhanced Safety of Vehicles* (2005).

[16] BÜCHSE, M., THIELE, M., AND MÜLLERSCHÖN, H. New developments on compression and transfer of simulation data within an sdm system. In *NAFEMS World Congress 2017 Summary of Proceedings* (2017), NAFEMS Ltd.

[17] BURCKEL, R. B., AND BAUER, H. *Probability Theory*. De Gruyter, 1995.

[18] CHAN, T. F., AND HANSEN, P. C. Computing truncated singular value decomposition least squares solutions by rank revealing qr-factorizations. *SIAM Journal on Scientific and Statistical Computing 11*, 3 (1990), 519–530.

[19] CHANDLER DAVIS, W. M. K. The rotation of eigenvectors by a perturbation. iii. *SIAM Journal on Numerical Analysis 7*, 1 (1970), 1–46.

[20] CICHOCKI, A., AND AMARI, S.-I. *Adaptive Blind Signal and Image Processing: Learning Algorithms and Applications*. John Wiley & Sons, Inc., New York, NY, USA, 2002.

[21] CLEARY, J. G., AND WITTEN, I. Data compression using adaptive coding and partial string matching. *Communications, IEEE Transactions on 32*, 4 (Apr 1984), 396–402.

[22] COOK, R. D., MALKUS, D. S., PLESHA, M. E., AND WITT, R. J. *Concepts and Applications of Finite Element Analysis*. John Wiley & Sons, 2007.

[23] CORMACK, G. V., AND HORSPOOL, R. N. Data compression using dynamic markov modelling. *The Computer Journal 30* (1986), 541–550.

[24] CORMEN, T. H., STEIN, C., RIVEST, R. L., AND LEISERSON, C. E. *Introduction to Algorithms*, 2nd ed. McGraw-Hill Higher Education, 2001.

[25] COVER, T. M., AND THOMAS, J. A. *Elements of information theory*, 2nd ed. Wiley-Interscience, New York, NY, USA, 2006.

[26] DASSAULT SYSTEMS. Abaqus. http://www.3ds.com/abaqus. Accessed on 2022-01-03.

[27] DAVIES, S., AND MOORE, A. Bayesian networks for lossless dataset compression. In *Proceedings of the fifth ACM SIGKDD international conference on Knowledge discovery and data mining* (1999), ACM, pp. 387–391.

[28] DAWES, B. boost c++ library. http://www.boost.org/. Accessed on 2022-01-03.

[29] DEVLIN, K. *The Joy of Sets: Fundamentals of Contemporary Set Theory*. Undergraduate Texts in Mathematics. Springer New York, 2012.

[30] DEVORE, R. A., JAWERTH, B., AND LUCIER, B. J. Image compression through wavelet transform coding. *IEEE Transactions on Information Theory 38*, 2 (March 1992), 719–746.

[31] DEVORE, R. A., JAWERTH, B., AND POPOV, V. Compression of wavelet decompositions. *American Journal of Mathematics 114*, 4 (1992), 737–785.

[32] DIESTEL, R. *Graphentheorie*. Springer Verlag, 2006.

[33] DÜBEN, P. D., LEUTBECHER, M., AND BAUER, P. New methods for data storage of model output from ensemble simulations. *Monthly Weather Review 147*, 2 (2019), 677–689.

[34] DUDA, J. Asymmetric numeral systems as close to capacity low state entropy coders. *CoRR abs/1311.2540* (2013).

[35] DURRETT, R. *Probability: Theory and Examples*. Cambridge Series in Statistical and Probabilistic Mathematics. Cambridge University Press, 2010.

[36] DYM, H. *Linear Algebra in Action*, vol. 78 of *Graduate Studies in Mathematics*. American Mathematical Society, Providence, Rhode Island, 2007.

[37] EKROOT, L., AND COVER, T. M. The entropy of markov trajectories. *IEEE Trans. Inf. Theor. 39*, 4 (Sept. 1993), 1418–1421.

[38] EMMRICH, E. *Gewöhnliche und Operator-Differentialgleichungen: Eine integrierte Einführung in Randwertprobleme und Evolutionsgleichungen für Studierende*. Vieweg-Studium : Mathematik. Vieweg+Teubner Verlag, 2004.

[39] ENGINEERING TOOLBOX. Speed of sound in some common solids. http://www.engineeringtoolbox.com/sound-speed-solids-d_713.html. Accessed on 2022-01-03.

[40] ESI GROUP. PAM-CRASH. https://www.esi-group.com/products/virtual-performance-solution. Accessed on 2022-01-03.

[41] EURO NCAP. European new car assessment programme. https://www.euroncap.com. Accessed on 2022-01-03.

[42] EVEN, S., AND EVEN, G. *Graph Algorithms*. Cambridge University Press, New York, NY, USA, 2011.

[43] FREY, B. J. *Bayesian networks for pattern classification, data compression, and channel coding*. PhD thesis, University of Toronto, 1997.

[44] GAILLY, J., AND ADLER, M. gzip 1.6. http://www.gzip.org. Accessed on 2022-01-03.

[45] GHAHRAMANI, Z. Hidden markov models. World Scientific Publishing Co., Inc., River Edge, NJ, USA, 2002, ch. An Introduction to Hidden Markov Models and Bayesian Networks, pp. 9–42.

[46] GNS MBH. Animator. https://gns-mbh.com/products/animator/. Accessed on 2022-01-03.

[47] GOLOMB, S. Run-length encodings (corresp.). *IEEE Transactions on Information Theory 12*, 3 (1966), 399–401.

[48] GOLUB, G. H., AND VAN LOAN, C. F. *Matrix Computations (3rd Ed.)*. Johns Hopkins University Press, Baltimore, MD, USA, 1996.

[49] GÖTSCHEL, S. *Adaptive Lossy Trajectory Compression for Optimal Control of Parabolic PDEs*. PhD thesis, Freie Universität Berlin, 2015.

[50] GÖTSCHEL, S., CHAMAKURI, N., KUNISCH, K., AND WEISER, M. Lossy compression in optimal control of cardiac defibrillation. *Journal of Scientific Computing 60*, 1 (2013), 35–59.

[51] GÖTSCHEL, S., AND WEISER, M. Lossy compression for pde-constrained optimization: adaptive error control. *Computational Optimization and Applications 62*, 1 (2015), 131–155.

[52] GÖTSCHEL, S., AND WEISER, M. Lossy compression for large scale pde problems. Tech. Rep. 19-32, ZIB, Takustr. 7, 14195 Berlin, 2019.

[53] GRAY, R. M. Conditional rate-distortion theory. Tech. rep., Stanford University, CA, Stanford Electronics Labs, 1972.

[54] GRAY, R. M. *Entropy and information theory*. Springer, New York, London, 2011.

[55] GRAY, R. M., AND DAVISSON, L. D. *An Introduction to Statistical Signal Processing*, 1st ed. Cambridge University Press, New York, NY, USA, 2010.

[56] GROSS, D., HAUGER, W., SCHRÖDER, J., AND WALL, W. *Technische Mechanik: Band 2: Elastostatik*. Springer-Lehrbuch. Springer Berlin Heidelberg, 2006.

[57] HALLQUIST, J. O. LS-DYNA Theory Manual. https://www.dynasupport.com/manuals/additional/ls-dyna-theory-manual-2005-beta/@@download/file/ls-dyna_theory_manual_2006.pdf, 2006. Accessed on 2022-01-03.

[58] HAUG, E. Engineering safety analysis via destructive numerical experiments. *EURO-MECH 12, Polish Academy of Sciences, Engineering Transactions 29*, 1 (1981), 39–49.

[59] HAUG, E., SCHARNHORST, T., AND DU BOIS, P. Fem-crash, berechnung eines fahrzeugfrontalaufpralls. *VDI Berichte 613* (1986), 479–505.

[60] HOEL, P. G., PORT, S. C., AND STONE, C. J. *Introduction to Stochastic Processes*. Houghton Mifflin Company, Boston, MA, 1972.

[61] HOTELLING, H. Analysis of a complex of statistical variables into principal components. *Journal of Educational Psychology 24*, 6 (1933), 417–441.

[62] IBARRIA, L., LINDSTROM, P., ROSSIGNAC, J., AND SZYMCZAK, A. Out-of-core compression and decompression of large n-dimensional scalar fields. *Computer Graphics Forum 22* (2003), 343–348.

[63] IBRAHIM, H. K. *Design optimization of vehicle structures for crashworthiness improvement*. PhD thesis, Concordia University, 2009.

[64] INTEL CORPORATION. Introduction to the intel mkl extended eigensolver. https://www.intel.com/content/www/us/en/develop/documentation/onemkl-developer-reference-fortran/top/extended-eigensolver-routines/the-feast-algorithm.html. Accessed on 2022-01-03.

[65] ISENBURG, M., AND ALLIEZ, P. Compressing polygon mesh geometry with parallelogram prediction. In *IEEE Visualization* (2002), pp. 141–146.

[66] ISING, E. Beitrag zur theorie des ferromagnetismus. *Zeitschrift für Physik A Hadrons and Nuclei 31*, 1 (1925), 253–258.

[67] JENSEN, F. V. *Bayesian Networks and Decision Graphs*. Springer-Verlag New York, Inc., Secaucus, NJ, USA, 2001.

[68] KAFSI, M., GROSSGLAUSER, M., AND THIRAN, P. The entropy of conditional markov trajectories. *IEEE Transactions on Information Theory 59*, 9 (2013), 5577–5583.

[69] KARHUNEN, K. Zur spektraltheorie stochastischer prozesse. *Annales Academiae Scientiarum Fennicae Series A1 - Mathematica Physica 34* (1946), 1–7.

[70] KARNI, Z., AND GOTSMAN, C. Compression of soft-body animation sequences. *Computers and Graphics 28* (2004), 25–34.

[71] KELLEY, C. T. *Iterative Methods for Linear and Nonlinear Equations*. SIAM, Philadelphia, PA, 1995.

[72] KIEFER, J. Sequential minimax search for a maximum. *Proceedings of the American Mathematical Society 4*, 3 (1953), 502–506.

[73] KLEIN, B. *FEM: Grundlagen und Anwendungen der Finite-Element-Methode im Maschinen-und Fahrzeugbau*. Springer-Verlag, 2010.

[74] KLENKE, A. *Wahrscheinlichkeitstheorie*. Springer, 2006.

[75] KÖNIGSBERGER, K. *Analysis 1:*. Analysis. Springer Berlin Heidelberg, 2003.

[76] KORB, K. B., AND NICHOLSON, A. E. *Bayesian Artificial Intelligence, Second Edition*, 2nd ed. CRC Press, Inc., Boca Raton, FL, USA, 2010.

[77] KUMAR, A., ZHU, X., TU, Y.-C., AND PANDIT, S. Compression in molecular simulation datasets. In *Intelligence Science and Big Data Engineering* (Berlin, 2013), vol. 8261 of *Lecture Notes in Computer Science*, Springer, pp. 22–29.

[78] LAURITZEN, S. L. *Graphical models*. Clarendon Press, 1996.

[79] LEE, J., AND VERLEYSEN, M. *Nonlinear dimensionality reduction*. Springer, New York, 2007.

[80] LENGYEL, J. E. Compression of time-dependent geometry. In *Proceedings of the 1999 Symposium on Interactive 3D Graphics* (1999), pp. 89–95.

[81] LI, J. *Image Classification and Compression Based on a two dimensional multiresolution hidden Markov model*. PhD thesis, Stanford University, 1999.

[82] LINDSTROM, P. Fixed-rate compressed floating-point arrays. *IEEE Transactions on Visualization and Computer Graphics 20*, 12 (Dec 2014), 2674–2683.

[83] LINDSTROM, P., AND ISENBURG, M. Fast and efficient compression of floating-point data. *IEEE Transactions on Visualization and Computer Graphics 12*, 5 (Sep. 2006), 1245–1250.

[84] LIVERMORE SOFTWARE TECHNOLOGY CORP. LS-DYNA. `http://www.lstc.com/products/ls-dyna`. Accessed on 2022-01-03.

[85] LIVERMORE SOFTWARE TECHNOLOGY CORP. LS-DYNA database binary output files. `http://www.dynamore.de/de/download/manuals/ls-dyna/ls-dyna-database-manual`, 2011. Accessed on 2022-01-03.

[86] LOVÁSZ, L. Random walks on graphs: A survey. In *Combinatorics, Paul Erdös is Eighty*, D. Miklós, V. T. Sós, and T. Szönyi, Eds., vol. 2. János Bolyai Mathematical Society, Budapest, 1996, pp. 353–398.

[87] MEI, L., AND THOLE, C. Uncertainty analysis for parallel car-crash simulation results. In *Current Trends in High Performance Computing and Its Applications*. Springer, 2005, pp. 393–398.

[88] MERTLER, S., AND MÜLLER, S. Reducing storage footprint and bandwidth requirements to a minimum: Compressing sets of. simulation results. In *LS-DYNA Forum 2016* (2016).

[89] MERTLER, S., MÜLLER, S., AND THOLE, C. Verfahren zur kompression von beobachtungen einer vielzahl von testabläufen, Sept. 30 2015. EP Patent App. EP20,150,158,584.

[90] MERTLER, S., MÜLLER, S., AND THOLE, C.-A. Predictive principal component analysis as a data compression core in a simulation data management system. In *Data Compression Conference (DCC), 2015* (April 2015), pp. 173–182.

[91] MEYWERK, M. *CAE-Methoden in der Fahrzeugtechnik*. Springer, 2007.

[92] MOFFAT, A., NEAL, R. M., AND WITTEN, I. H. Arithmetic coding revisited. *ACM Trans. Inf. Syst. 16*, 3 (July 1998), 256–294.

[93] MOFFAT, A., AND TURPIN, A. *Compression and coding algorithms / by Alistair Moffat and Andrew Turpin*. Kluwer Academic Publishers Boston ; London, 2002.

[94] MÜLLER, K., SMOLIC, A., KAUTZNER, M., EISERT, P., AND WIEGAND, T. Predictive compression of dynamic 3d meshes. In *Proceedings of ICIP 2005* (2005), vol. 1, pp. 621–624.

[95] MÜLLER, S. Hochpass-Quantisierung für die Kompression von Simulationsergebnissen. Diplomarbeit, Universität zu Köln, 2010.

[96] MÜLLER, S. P. Induced markov chains for the encoding of graph-based data. In *Signal Processing and Information Technology (ISSPIT), 2014 IEEE International Symposium on* (Dec 2014), pp. 143–148.

[97] MÜLLER, S. P., AND BANERJEE, P. S. FEMZIP-CFD: Taming of the data pile. In *NAFEMS Technical Paper* (2015), NAFEMS.

[98] NATIONAL HIGHWAY TRAFFIC SAFETY ADMINISTRATION. Cheverolet Silverado. http://www-nrd.nhtsa.dot.gov/Departments/Crashworthiness/ Crashworthiness%20by%20vehicle%20Models/LSDYNA_FE_MODELS/SILVERADO/ Silverado.zip. Accessed on 2022-01-03.

[99] NAVARD, S. E., SEAMAN, J. W., AND YOUNG, D. M. A characterization of discrete unimodality with applications to variance upper bounds. *Annals of the Institute of Statistical Mathematics 45*, 4 (1993), 603–614.

[100] NORRIS, J. R. *Markov chains*. Cambridge series in statistical and probabilistic mathematics. Cambridge University Press, 1998.

[101] PEARL, J., AND PAZ, A. *Graphoids: A graph-based logic for reasoning about relevance relations*. University of California (Los Angeles). Computer Science Department, 1985.

[102] PEARSON, K. On lines and planes of closest fit to systems of points in space. *Philosophical Magazine 2*, 6 (1901), 559–572.

[103] POLIZZI, E. Density-matrix-based algorithm for solving eigenvalue problems. *Phys. Rev. B 79* (Mar 2009), 115112.

[104] REDDY, J. N. *An Introduction to Nonlinear Finite Element Analysis: with applications to heat transfer, fluid mechanics, and solid mechanics*. OUP Oxford, 2014.

[105] REINHARDT, H. *Aufgabensammlung Analysis 1: mit mehr als 500 Übungen und Lösungen*. Springer Berlin Heidelberg, 2016.

[106] RETTENMEIER, M. Zwei Strategien zur verlustfreien Kompression von Simulationsergebnissen. Diplomarbeit, Universität zu Köln, 2007.

[107] RETTENMEIER, M. *Data Compression for Computational Fluid Dynamics on Irregular Grids*. Logos Verlag, 2012.

[108] REYES, M. *Cutset Based Processing and Compression of Markov Random Fields*. PhD thesis, University of Michigan, 2010.

[109] REYES, M. G. personal communication, February 2014.

[110] RICE, R. F. Some practical universal noiseless coding techniques. In *Proceedings of the SPIE Symposium* (1979), vol. 207.

[111] ROELOFS, G., GAILLY, J., AND ADLER, M. zlib 1.2.8. http://www.zlib.net. Accessed on 2022-01-03.

[112] ROMASZEWSKI, M., GAWRON, P., AND OPOZDA, S. Dimensionality reduction of dynamic animations using ho-svd. In *Artificial Intelligence and Soft Computing*, vol. 8467 of *Lecture Notes in Computer Science*. Springer International Publishing, 2014, pp. 757–768.

[113] SAAD, Y. *Iterative Methods for Sparse Linear Systems*, 2nd ed. Society for Industrial and Applied Mathematics, Philadelphia, PA, USA, 2003.

[114] SAID, A. Comparative analysis of arithmetic coding computational complexity. In *Data Compression Conference, 2004. Proceedings. DCC 2004* (March 2004), pp. 562–.

[115] SALOMON, D. *A guide to data compression methods*. Springer, New York, NY, 2001.

[116] SALOMON, D. *Data Compression: The Complete Reference*. Springer-Verlag, Berlin, Germany, 2007.

[117] SATTLER, M., SARLETTE, R., AND KLEIN, R. Simple and efficient compression of animation sequences. In *Proceedings of the 2005 ACM SIGGRAPH/Eurographics Symposium on Computer Animation* (New York, NY, USA, 2005), SCA '05, ACM, pp. 209–217.

[118] SCALE GMBH. LoCo. https://www.scale.eu/de/produkte/loco. Accessed on 2022-01-03.

[119] SCHILLING, R. *Measures, Integrals and Martingales*. No. Bd. 13 in Measures, integrals and martingales. Cambridge University Press, 2005.

[120] SCHWARZ, H., AND KÖCKLER, N. *Numerische Mathematik*. Lehrbuch Mathematik. Vieweg+Teubner Verlag, 2007.

[121] SHANNON, C. E. A mathematical theory of communication. *Bell system technical journal 27* (1948).

[122] SHAWE-TAYLOR, J., WILLIAMS, C., CRISTIANINI, N., AND KANDOLA, J. On the eigenspectrum of the gram matrix and its relationship to the operator eigenspectrum. In *Algorithmic Learning Theory*, N. Cesa-Bianchi, M. Numao, and R. Reischuk, Eds., vol. 2533 of *Lecture Notes in Computer Science*. Springer Berlin Heidelberg, 2002, pp. 23–40.

[123] SIDACT GMBH. DIFFCRASH. https://www.sidact.de/diffcrash. Accessed on 2022-01-03.

[124] SIDACT GMBH. FEMZIP Crash. https://www.sidact.de/femzip. Accessed on 2022-01-03.

[125] SORKINE, O., COHEN-OR, D., AND TOLEDO, S. High-pass quantization for mesh encoding. In *Proceedings of the 2003 Eurographics/ACM SIGGRAPH Symposium on Geometry Processing* (Aire-la-Ville, Switzerland, Switzerland, 2003), SGP '03, Eurographics Association, pp. 42–51.

[126] STEFFES-LAI, D. *Approximation Methods for High Dimensional Simulation results - Parameter Sensitivity Analysis and Propagation of Variations for Process Chains.* Logos Verlag Berlin, 2014.

[127] STEWART, W. *Introduction to the numerical solution of Markov chains.* Princeton Univ. Press, Princeton, NJ, 1994.

[128] STEWART, W. *Probability, Markov Chains, Queues, and Simulation: The Mathematical Basis of Performance Modeling.* Princeton University Press, 2009.

[129] STOER, J., AND BULIRSCH, R. *Numerische Mathematik 2,* 5 ed. Grundkurs Mathematik. Springer, Berlin, 2005.

[130] STROOCK, D. W. *An introduction to Markov processes.* Graduate texts in mathematics. Springer, Berlin, 2005.

[131] TAUBIN, G., AND ROSSIGNAC, J. Geometric compression through topological surgery. *ACM Trans. Graph. 17,* 2 (Apr. 1998), 84–115.

[132] THOLE, C.-A. Compression of ls-dyna3d simulation results using femzip. In *3. LS-DYNA Anwenderforum* (Bamberg, 2004), pp. E–III–1–5.

[133] THOLE, C.-A. New developments in the compression of ls-dyna simulation results using femzip. In *6th European LS-DYNA Users' Conference* (Bamberg, 2007), pp. 3.69–3.76.

[134] THOLE, C.-A. Advanced mode analysis for crash simulation results. In *11th International LS-DYNA Users Conference* (Sankt Augustin, Germany, 2010).

[135] TOUMA, C., AND GOTSMAN, C. Triangle mesh compression. *PROC GRAPHICS INTERFACE. pp. 26-34. 1998* (1998).

[136] TRENDELKAMP-SCHROER, B. personal communication, March 2014.

[137] VASA, L. Optimised mesh traversal for dynamic mesh compression. *Graphical Models 73,* 5 (2011), 218 – 230.

[138] VASA, L., AND SKALA, V. Coddyac: Connectivity driven dynamic mesh compression. In *3DTV Conference, 2007* (May 2007), pp. 1–4.

[139] VASA, L., AND SKALA, V. Cobra: Compression of the basis for pca represented animations. *Computer Graphics Forum 28,* 6 (2009), 1529–1540.

[140] WEISER, M. personal communication, May 2017.

[141] WEISER, M., AND GÖTSCHEL, S. State trajectory compression for optimal control with parabolic pdes. *SIAM J. Sci. Comput. 34,* 1 (Jan. 2012), 161–184.

[142] WIDROW, B., AND KOLLÁR, I. *Quantization Noise: Roundoff Error in Digital Computation, Signal Processing, Control, and Communications.* Cambridge University Press, Cambridge, UK, 2008.

[143] WILLEMS, F. M., AND TJALKENS, T. J. Chapter 1 - information theory behind source coding. In *Lossless Compression Handbook*, K. Sayood, Ed., Communications, Networking and Multimedia. Academic Press, San Diego, 2003, pp. 3 – 34.

[144] WINTER, R., MANTUS, M., AND PIFKO, A. B. Finite element crash analysis of a rear-engine automobile. Tech. rep., SAE Technical Paper, 1981.

[145] YEH, P.-S. Chapter 16 - the {CCSDS} lossless data compression recommendation for space applications. In *Lossless Compression Handbook*, K. Sayood, Ed., Communications, Networking and Multimedia. Academic Press, San Diego, 2003, pp. 311 – 326.

[146] YEUNG, R. W. *A First Course in Information Theory (Information Technology: Transmission, Processing and Storage).* Springer-Verlag New York, Inc., Secaucus, NJ, USA, 2006.

[147] ZIENKIEWICZ, O. C., AND TAYLOR, R. L. *The finite element method: solid mechanics,* vol. 2. Butterworth-Heinemann, 2000.

[148] ZIENKIEWICZ, O. C., AND TAYLOR, R. L. *The finite element method: the basis,* vol. 5. Butterworth-Heinemann, 2000.

[149] ZIENKIEWICZ, O. C., TAYLOR, R. L., AND ZHU, J. Z. *The Finite Element Method: Its Basis and Fundamentals, Sixth Edition,* 6 ed. Butterworth-Heinemann, May 2005.

[150] ZIV, J., AND LEMPEL, A. A universal algorithm for sequential data compression. *IEEE Transactions on Information Theory 23,* 3 (May 1977), 337–343.